Environmental Ethics: A Very Short Introduction

VERY SHORT INTRODUCTIONS are for anyone wanting a stimulating and accessible way into a new subject. They are written by experts, and have been translated into more than 45 different languages.

The series began in 1995, and now covers a wide variety of topics in every discipline. The VSI library currently contains over 550 volumes—a Very Short Introduction to everything from Psychology and Philosophy of Science to American History and Relativity—and continues to grow in every subject area.

Very Short Introductions available now:

ABOLITIONISM Richard S. Newman
ACCOUNTING Christopher Nobes
ADAM SMITH Christopher J. Berry
ADOLESCENCE Peter K. Smith
ADVERTISING Winston Fletcher
AFRICAN AMERICAN RELIGION
    Eddie S. Glaude Jr
AFRICAN HISTORY John Parker and
    Richard Rathbone
AFRICAN POLITICS Ian Taylor
AFRICAN RELIGIONS
    Jacob K. Olupona
AGEING Nancy A. Pachana
AGNOSTICISM Robin Le Poidevin
AGRICULTURE Paul Brassley and
    Richard Soffe
ALEXANDER THE GREAT
    Hugh Bowden
ALGEBRA Peter M. Higgins
AMERICAN CULTURAL HISTORY
    Eric Avila
AMERICAN HISTORY Paul S. Boyer
AMERICAN IMMIGRATION
    David A. Gerber
AMERICAN LEGAL HISTORY
    G. Edward White
AMERICAN POLITICAL HISTORY
    Donald Critchlow
AMERICAN POLITICAL PARTIES
    AND ELECTIONS L. Sandy Maisel
AMERICAN POLITICS
    Richard M. Valelly
THE AMERICAN PRESIDENCY
    Charles O. Jones

THE AMERICAN REVOLUTION
    Robert J. Allison
AMERICAN SLAVERY
    Heather Andrea Williams
THE AMERICAN WEST Stephen Aron
AMERICAN WOMEN'S HISTORY
    Susan Ware
ANAESTHESIA Aidan O'Donnell
ANALYTIC PHILOSOPHY
    Michael Beaney
ANARCHISM Colin Ward
ANCIENT ASSYRIA Karen Radner
ANCIENT EGYPT Ian Shaw
ANCIENT EGYPTIAN ART AND
    ARCHITECTURE Christina Riggs
ANCIENT GREECE Paul Cartledge
THE ANCIENT NEAR EAST
    Amanda H. Podany
ANCIENT PHILOSOPHY Julia Annas
ANCIENT WARFARE
    Harry Sidebottom
ANGELS David Albert Jones
ANGLICANISM Mark Chapman
THE ANGLO-SAXON AGE John Blair
ANIMAL BEHAVIOUR
    Tristram D. Wyatt
THE ANIMAL KINGDOM
    Peter Holland
ANIMAL RIGHTS David DeGrazia
THE ANTARCTIC Klaus Dodds
ANTHROPOCENE Erle C. Ellis
ANTISEMITISM Steven Beller
ANXIETY Daniel Freeman and
    Jason Freeman

Robin Attfield

ENVIRONMENTAL ETHICS

A Very Short Introduction

OXFORD
UNIVERSITY PRESS

OXFORD
UNIVERSITY PRESS

Great Clarendon Street, Oxford, OX2 6DP,
United Kingdom

Oxford University Press is a department of the University of Oxford.
It furthers the University's objective of excellence in research, scholarship,
and education by publishing worldwide. Oxford is a registered trade mark of
Oxford University Press in the UK and in certain other countries

Published in the United States of America by Oxford University Press
198 Madison Avenue, New York, NY 10016, United States of America

British Library Cataloguing in Publication Data
Data available

Library of Congress Control Number: 2018946418

ISBN 978-0-19-879716-6

Printed in Great Britain by
Ashford Colour Press Ltd., Gosport, Hampshire.

Contents

Contents

Acknowledgements

Thanks are due to Cardiff University Institute for Sustainable Places for facilitating the composition of this book, and to one of their visiting speakers, Hilary Graham, for bibliographical assistance relating to the work of her research team. Likewise I would like to thank Jonathan Helfand for help in tracking down a chapter of his, and to the staff of Oxford University Press, and to Jenny Nugée in particular, for assiduous assistance with many aspects of the preparation of this book. Thanks are also due to Matthew Quinn of Sustainable Places for commenting on drafts of some chapters and to the Cardiff University technicians for sorting computer-related glitches.

Special thanks are due to two anonymous OUP readers, one for comments as first drafts of the successive chapters emerged, and the other for comments on the manuscript as a whole and for several suggested paragraph-length redrafts, some of which have been adjusted and adopted. I am also grateful to the authors of the endorsements (which appear on the back cover of the book).

Thanks go too to the people with whom I have written joint papers during the period when this book was emerging: Rebekah Humphreys, Melissa Beattie, and Kate Attfield. As ever, my biggest debt is to my wife, Leela Dutt Attfield, without whom this entire project would have been inconceivable.

List of illustrations

Chapter 1
Origins

Environmental problems

Nature is disappearing fast, or so we are led to believe. Fewer whales swim the oceans. Fewer tigers stalk the Sundarbans of Bengal. Many coral reefs are bleaching, putting their polychrome communities at risk. The habitats of orang-utans in Sumatra and Borneo are threatened. Freak hurricanes blight the Caribbean and shred its trees. Closer at hand, garden birds and butterflies are dwindling in number. In Britain, even bluebells and Wordsworth's wild daffodils are said to be endangered. What, we may wonder, is going on?

The natural world has long ceased to be a reliable backdrop to human life, unaffected by human activity. For many centuries we have been changing it, through hunting and farming, through building, mining, and engineering, and through travelling and trading. We may still think of it as our unceasing, enduring environment, unchanging as the stars above us, but the environment that our grand-children inherit will be vastly different from that of our early ancestors, and even from the environment we were born into ourselves. We can no longer take it for granted, even if we ever could.

Because of human impacts on the world of nature, many people call the present age 'the Anthropocene', coining this term to echo geological ages such as the Eocene and the Pleistocene. What they mean is that human impacts have become predominant over the whole surface of the Earth.

They fail to agree about when this age began. Did it begin with the invention of ships, with the industrial revolution, or with the world wars of the 20th century? There is no agreement either on whether this means that it is too late to preserve the natural world, whether we are free to remould the face of the Earth as we please (for a version of this view, see the section of Chapter 6 on social ecology), or whether we should use our knowledge and technology to restore tracts of the world to their pre-human

1. Our planet, as seen from the depths of space (courtesy of NASA). We have no other.

condition. But they agree that humankind has become one of the main influences on the face of our planet. (See Figure 1.)

Deforestation and soil erosion are among ways in which people have changed the natural world. Alongside positive developments such as the building of cities, others include the loss of numerous species, the growth of deserts, the depletion of resources, pollution, and, as we have discovered in recent decades, climate change. These processes used not to be regarded as environmental problems, because nature appeared inexhaustible. Problems only come to be recognized as such when they are seen to involve avoidable harms, and when ways can be envisaged to solve or at least alleviate them, as John Passmore has sagely remarked in *Man's Responsibility for Nature*.

Among philosophers, Plato (in his dialogue *Critias*) was one of the earliest to be aware of soil erosion and deforestation, but he was untroubled by these developments, as was his disciple Aristotle, who, in his *Meteorologica*, depicted nature as permanent and fundamentally unchanging. It was not until the 19th century that people like George Perkins Marsh, in *Man and Nature* (1864), came to regard nature as significantly vulnerable to human activity, and at the same time human life as vulnerable to nature and its changes.

The 20th century saw the rise of ecological science, and the related study of nature as composed of interacting natural systems, but the case for preserving systems such as rivers and forests had to await the publication of Aldo Leopold's *A Sand County Almanac* (1949). Leopold advocated extending ethics to encompass ecosystems, but philosophers and ethicists (Leopold was neither of these) remained unimpressed. What may have served to change the atmosphere was Rachel Carson's work *Silent Spring* (1962), with its disclosure that pesticides such as DDT (dichlorodiphenyltrichloroethane), used in Europe, were now to be found in the flesh of Antarctic penguins.

Another factor was the spectacle of defoliation used by American forces during the American intervention in Vietnam (1961–75), with its implicit attempt to embark on biological warfare and to sequestrate or even eradicate the natural world of central Indo-China. The new awareness of the unexpected side-effects of human impacts on the environment, and how human action can imperil whole species and ecosystems, emboldened ethicists to redirect their focus to environmental issues.

The emergence of environmental ethics

Philosophical ethics had for some decades held back (at least in the Anglo-Saxon world) from reflection on practical issues, focusing instead on the analysis and the meaning of concepts. But from the 1960s new issues in medicine (such as experimentation on human subjects and the requirements of informed consent) brought a new lease of life to the ancient sub-discipline of medical ethics, and the spread of nuclear weapons rekindled reflection on the ethics of war.

The stage was thus set for the emergence in the early 1970s of environmental philosophy and ethics, and related attempts to apply philosophy to environmental concepts and problems. Up to the start of the 20th century, philosophy had always been understood as applicable to practical issues (think of the political philosophy of Plato, Aristotle, Spinoza, Locke, and Kant). The various branches of applied philosophy now set about rescuing this longstanding tradition and bringing it back to life and vigour.

At a World Congress of Philosophy held in Bulgaria in 1973, Richard Routley (later Sylvan), an Australian philosopher, gave an address entitled 'Is There a Need for a New, an Environmental Ethic?' His answer to this question was emphatically affirmative. He took the traditional Western view to be that only human interests matter, and that we humans may treat nature as we please. He rejected this view on the basis of thought-experiments.

For example, if 'the Last Man', a survivor of a nuclear holocaust, lays about him, eliminating, as far as he can, every remaining living being, animal or vegetable, what he does would be permissible for the traditional view, but in most people's intuitive judgement his action is to be condemned as wrong. Such thought-experiments (several were presented) disclose, Routley argued, that there is a growing environmental ethic at odds with the traditional view, and one which better responds to the assaults of human beings on the natural world. We should thus reject the human-interests-only stance (soon to be called 'anthropocentrism'), and adopt a stance for which other living creatures matter as well.

One widespread response to Routley's thought-experiments is that they concern such extreme and exceptional circumstances that people's intuitive judgements about them cease to be reliable, let alone indicative of the principles that we need. Critics suggested that, when judging the deeds of the Last Man, we inadvertently smuggle back into the scenario assumptions that fit more normal cases. We assume (they say) that other people or future people will somehow suffer from his behaviour, even though Routley's scenario was devised specifically to exclude all this.

Yet Routley could reply that he needs to supply a scenario of this kind to allow us to make judgements about a case where there are no remaining human interests (the Last Man, we may imagine, is shortly going to die himself), and where the only interests at stake are those of non-human animals and plants. Besides, he could insist that even in cases where it is clear that no human interests remain at stake, most people still consider it wrong to destroy other living beings.

So Routley's argument against anthropocentrism and in support of a new environmental ethic was widely found to be persuasive. At the very least it seemed to show that non-human animals should be taken into consideration in human decision-making. And if his thought-experiment were adjusted to exclude the remaining presence of animal interests (if, say, all animals in the vicinity had been killed by the same nuclear holocaust), the widespread judgement that the Last Man would be acting wrongly in destroying, as far as he could, the surviving plants could be held to suggest that the good of plants should be regarded as mattering, from an ethical perspective, alongside that of non-human animals and human beings.

But was Routley right in his characterization of Western traditions? He was responding to a depiction of the Western tradition by John Passmore, whose book *Man's Responsibility for Nature* was published the following year (1974). Passmore held that the majority view was human-centred and involved no ethical restrictions on the treatment of nature.

Yet he also recognized two minority traditions. In one of these, human beings are stewards or trustees of the world of nature, and responsible for its care (hence the title of his book, *Man's Responsibility for Nature*)—and, in religious versions of this tradition, answerable for their stewardship to God. In the second tradition, the role of human beings is to enhance or perfect the world of nature by cooperating with and bringing out its potential. Both these 'minority' traditions were held to have ancient roots and a long history in Western culture, and thus Passmore's suggestion was that the development of an environmental ethic need not involve a complete rejection of these traditions, which are richer than is often recognized, but can rather involve moving towards these other traditional stances.

For his part, Routley maintained that Passmore's 'minority' views were fundamentally human-centred themselves, and, because they supposedly fail to take into account non-human interests, need to be rejected and superseded. But these claims can be contested; for there is evidence that both of Passmore's 'minority' traditions were widely held and advocated in the early centuries of Christianity, and are thus hardly minority traditions at all. Equally, they can be interpreted (and have long been interpreted) in ways that recognize the ethical importance of non-human interests as well as the interests of human beings. (Much of this evidence had already been assembled by Clarence Glacken: see Chapter 7.)

Routley's contribution, then, was an important one with regard to the kind of ethic required, but his narrow view of Western traditions and their resources needs to be taken with a considerable pinch of salt. Many of the saints, for example, were prominent in treating animals, both wild and domesticated, with concern and kindness; so a broader view of Western traditions could well be preferable.

Naess and Deep Ecology

In the same year as Routley's World Congress address, the philosophical journal *Inquiry* published another ground-breaking paper, this one by the Norwegian philosopher Arne Naess, 'The Shallow and the Deep, Long-Range Ecology Movement. A Summary'. Naess contrasted two kinds of ecology movement.

The shallow kind is concerned with human interests of the next fifty years or so, and in particular with those of the people of developed countries. By contrast, the deep kind is additionally concerned with the good of the people of developing countries, with the long-term future, and with non-human species, affirming their 'equal right to live and blossom'. Naess recognized the practical need for some harvesting and killing of animals and plants if human life was to continue, but still adhered in principle to what he called 'biospherical egalitarianism' or the equal entitlement of all species to live their own way of life.

Naess's advocacy of the Deep Ecology movement involved support for a broad platform of stances (for such broad inclusiveness is part of what he meant by a 'movement'), including biological diversity. (By 'biological diversity' he will have intended promoting or preserving as full a range as possible of species, sub-species, and habitats.) He regarded the cultivation of such diversity as life-enhancing, probably having in mind its fostering of non-human life and enriching human life simultaneously.

At the same time cultural diversity was commended too, together with opposition to inter-human oppression such as exploitation through economic advantage or the power of class. Pollution and resource-depletion were to be contested, not just local forms affecting developed countries, or ones ignoring wider, global perspectives. The central value was self-realization, or the fulfilment of the potentials of organisms of every kind.

While others might focus on different values from those just mentioned, Naess's approach has much to offer, not least its stress on self-realization and its globally inclusive scope. But his 'Deep Ecology' platform also includes some controversial claims, including his account of personal identity.

For Naess, my true self is not confined to my physical body, but (because everything is connected to everything else) extends to the whole of nature. It is this extended or greater Self that I am (supposedly) obliged to defend. But this move seems to take identification far too far. Besides, many people find that tracts of nature are well worth defending even if they do not identify with nature in this way. For there are plenty of other motivations, such as respect, admiration, and wanting our successors to be free to appreciate the same scenes as ourselves.

More worryingly, the 'Deep Ecology' platform advocated a significant reduction of the human population. This tenet was thought to be needed to allow room for the continued flourishing of other species. But it also led some of Naess's followers (though not Naess himself) to welcome catastrophes like famines, and the consequent decrease in human numbers they were expected to bring. Others were inclined to reject any platform capable of carrying such implications. We should perhaps respond to the 'Deep Ecology' platform cautiously and selectively.

Rolston's contribution

Another striking foundational contribution to environmental ethics was Holmes Rolston III's early essay 'Is There an Ecological Ethic?' (1975). Rolston (an American philosopher, now widely regarded as the father of environmental philosophy) was concerned to explain why we ought, for example, to recycle, and thus how to get from facts and scientific laws to conclusions that hinge around an 'ought', particularly those of an ecological kind.

The problem of justifying statements with an 'ought' at their centre was a longstanding one, drawn to attention in the 18th century by the philosopher David Hume. But Rolston was able to suggest more than one solution for the recycling example he had selected. A first possible solution is that (ultimately) human life depends on recycling (through life-supporting ecosystems being maintained), and that human life is itself valuable. This approach makes recycling a matter of fostering human interests, and (Rolston would say) embodies humanist rather than ecological values. However, Rolston's preferred solution instead says that we should recycle because this promotes ecosystem integrity, and ecosystem integrity has value in itself, or intrinsic value.

Much could be said to elucidate the notion of ecosystem integrity. It may suffice to say that this would involve healthy, functioning ecosystems, both incorporating and supporting interacting living organisms and their cycles of life. Here, Rolston's thinking evokes echoes of Leopold's earlier advocacy of maintaining the integrity, stability, and beauty of the biosphere.

But Rolston was also drawing attention to the need for environmental ethics to adopt an understanding of value that does not stop short at what is valuable merely as a means (like money

9

and resources), and instead goes on to identify what is valuable for its own sake. A fairly uncontroversial example of something valuable for its own sake is health.

Nearly everyone takes it for granted that something or other is valuable for its own sake. Very few people seriously believe that nothing at all has this character. Rolston's distinctive suggestion is that an ecological ethic might stand out from other approaches to ethics through finding what is fundamentally valuable not only in human fulfilments, but also in non-human lives or well-being, or perhaps in the biological systems of which they are part.

One common feature of the contributions of Routley, Naess, and Rolston was their rejection of a human-interests-only or 'anthropocentric' approach to ethics. Here many readers may wish to sympathize, at least tentatively.

Yet the issue soon arose of whether you can study environmental ethics at all if you endorse such an anthropocentric approach, or whether you are disqualified before you begin. But subjects of study should not be defined by ideological stances, and in any case one of the reasons for environmentally friendly activities like recycling could well be that they benefit human beings.

So, while many of us may wish to support one or another ampler and broader value-theory than anthropocentrism, it would be wise not to banish anthropocentrist thinkers from the community of environmental ethicists, and certainly not to exclude them by definition. Such thinkers often call themselves 'pragmatists', and some, such as Bryan Norton, have actually made important contributions to this field.

Themes and issues arising

One issue raised by these early thinkers was the question of which beings matter where ethics is concerned, and should be taken into account when decisions are being made. To use different language, this is the question of the scope of moral standing.

One cogent answer to this question has been supplied by Kenneth Goodpaster: whatever has a good of its own and can be benefited. For bestowing benefits is central to morality. In other words, all living organisms have moral standing. Stances of this kind have been called 'biocentric', in emphatic contrast to the 'anthropocentric' approach of some traditional views. Another kind of answer will be mentioned shortly.

A further issue concerns whether, and how much, future interests count. (Naess's essay in particular raises this matter.) Aristotle thought that including these interests would make ethics too complicated. But the effects of modern technology are often foreseeable, and it would be irresponsible to utilize this technology while ignoring them. So the future impacts of current actions should be taken into account where they can be foreseen. In this context, Hans Jonas has argued (in *The Imperative of Responsibility*) that the newly enlarged range of impacts of human behaviour on future generations and on non-human species requires reconfiguring our conception of ethics itself.

Yet most economists believe in discounting future goods and bads so that they count for less than present ones. They have some good reasons, because (for example) some future impacts are uncertain. But philosophers have tended to respond that discounting should be limited to cases where these reasons can be seen to apply, and not applied across the board, or blanketwise. Future injuries and future pollution will be just as bad as present cases, and, when foreseeable, should be treated just as seriously.

Another issue, raised seminally by Rolston (and touched on already in the previous section of this chapter), concerns what has value not as a means (or instrumentally) but in itself (or intrinsically). Things of this kind will be what give moral 'oughts' their point. One possible answer is the flourishing of human beings. But if we accept that other living creatures also have moral standing, then the flourishing of these creatures must be seen as having intrinsic value as well.

Some philosophers have reservations about the very concept of intrinsic value. But if anything has value of any other sort, then something must have intrinsic value (value that is not derivative from something else). For if nothing had such value, then nothing would have value at all. And while human happiness or flourishing is usually agreed to be one such 'something', reflection on the themes discussed in this chapter (for example on Routley's Last Man thought-experiment) suggests that most of us assume that the flourishing of other creatures is another such 'something'.

One final issue concerns the question of the grounds for preserving species and ecosystems. Some environmental ethicists suggest that these entities must be held to have intrinsic value themselves, a stance called 'ecocentrism'. We do seem to attach greater importance to preserving the last members of a species than (like numbers of) members of unthreatened species; and this tendency could be predicted if ecocentrism is correct. But it could instead be due to the moral standing and intrinsic value of future species-members, the existence of which depends on the survival of current ones. So biocentrism can answer this question too. Similar reasoning is relevant to ecosystems.

Anthropocentrism can also answer this question, but only for species and ecosystems that are beneficial to humanity. Arguably, though, many are not. Does this mean that there is no case for preserving them? The answer to this question may allow you, the reader, to discover where you stand yourself in matters of environmental preservation.

Chapter 2
Some key concepts

Nature

In this chapter, the focus is on key concepts, widely held pivotal to thinking about environmental ethics, as well as other ethical fields. I begin with the concept of nature, ways in which people seek to relate nature to human behaviour, and attitudes both to the nature that surrounds us and to our inner nature too.

Are human beings apart from nature or simply part of nature? If we are simply part of nature, we could reason (as some people do) that whatever we do is natural, and (they sometimes add) therefore beyond criticism. But this would make ethics redundant (and environmental ethics too), for whatever we may do would be both natural and right.

However, if human beings are distinct from nature, it seems to follow that we cannot have evolved from natural creatures, and that they are not our kin (contrary to Darwinism). It may seem to follow that nature is an enemy to be overcome. It even seems to follow that we do not have a nature, and may be moulded, with no harm done, into whatever way of life the authorities may prefer (as totalitarians sometimes claim).

To avoid these unwelcome apparent implications, we need to clarify the concept of nature. Thus if 'nature' or 'natural' means whatever is not supernatural, then human beings are clearly natural. But this does not mean that their behaviour is exempt from ethical standards, nor beyond criticism. Human behaviour might be beyond criticism if it were in all respects biologically determined, but that would bring in quite a different (and a highly questionable) sense of 'natural', not implicit in being natural in the sense of 'non-supernatural', and one that would need to be argued for rather than assumed.

Another sense of 'natural' (and, like the previous one, highlighted by John Stuart Mill) contrasts 'natural' with 'artificial'; what is natural is not significantly affected by human choices or culture. In this sense, tropical rainforests and spiral nebulae may be natural, but cities and motorways are not, and neither is art, cookery, or sport. Much human life will not be natural (in this sense), because most people have human upbringings and education.

But nothing follows about human beings not having evolved from or not being dependent on other organisms. Nor does it follow that we have no inherited nature, or that we cannot be harmed by authoritarian attempts to mould our lives. Much less does it follow that nature is something that human beings should conquer or subdue; being ourselves dependent on natural (or non-artificial) organisms and forces, we would, by pursuing this goal, be striking suicidally at systems on which our own lives and our children's depend.

Our having or not having a 'nature' brings in yet another sense of 'nature', not related to being either non-supernatural or to being non-artificial. In this sense, our nature is our make-up or what it is that makes us what we are, and what is natural consists in the characteristics that this involves. Our well-being depends on not being subjected to unnatural factors, such as excessive stress, where 'unnatural' contrasts with this sense of 'natural'. Since this is a different sense of 'nature' from the others, no one can argue that, because little of human life is natural (in the sense of non-artificial), human beings lack natures or can be treated just anyhow on this basis.

So we need to be as clear as we can which sense of 'nature' or 'natural' we are using. The weird conclusions of the second and third paragraphs above are only reached by switching senses in mid-stream. Like everyone else, environmental ethicists (and readers of this book too) need to avoid confusing or conflating different senses. That is the way to avoid (so to speak) 'cruel and unnatural' conclusions, not least about nature.

The suggestion is sometimes made that what is valuable and to be aimed at is simply what is natural, in the sense of non-artificial. This suggestion has the merit of finding value in non-human nature, and thus avoiding anthropocentrism. But at the same time it fails to distinguish between living creatures on the one hand and inanimate entities on the other; for it is implausible that the latter (rocks and the like) have value on the same basis as plants and animals. It also appears to disown the value of human art, workmanship, and creativity, for all of these turn on human culture and artifice. Accordingly a more discriminating understanding of value seems to be needed. This is returned to later in this chapter.

But it should be added that when people talk of 'nature' (and threats to it) they often mean 'wildlife'. They have in mind issues like the endangered status of species such as the turtle-dove and the corncrake in Britain. (See Figure 2.) Here specific explanations, such as the use of agro-chemicals, can be suggested; if so, better farming methods may offer a remedy. But many problems of species extinction or attenuation are ascribable to global warming, an issue to be discussed in Chapter 8. As long as the sense in which 'nature' is being used is clear, then relevant problems can begin to be understood and addressed.

2. *Agalychnis annae*: an endangered tree-frog from Costa Rica; a threatened species on the IUCN Red List. Imagine a world without them.

It is sometimes suggested, however, that we ought to follow nature or live in accordance with it. The usual suggestion here is partly to imitate the processes of nature, and partly to follow our instincts and genetic tendencies, with as little artifice as possible.

John Stuart Mill gave a scathing reply to suggestions of this sort. The operation of nature is often merciless and unrelenting, especially towards sick and vulnerable creatures, and it is a great achievement of human civilization to have abandoned such a way of life.

It could be added that desirable social policies of the kind that many environmentalists would welcome, such as ecological education and conservation, depend on the exercise of conscious human choice, and thus on *not* letting nature take its course. When it comes to art, there is often a case for imitating nature, and when it comes to pest control, there is often a case for borrowing or imitating natural processes rather than using ecologically disruptive chemicals. But in matters of general ethical guidance, following nature is seldom the best course.

Ancient Stoicism was an influential ethical system which advised humanity to 'follow Nature'. But since nature as a whole is such a vast degree greater and more powerful than ourselves, the Stoics concluded that what 'following Nature' turns out to mean in practical terms is ceasing to try to control the events in our lives and focusing instead on controlling our feelings about them. For the events of our lives (as opposed to our feelings) were supposed to be determined by nature, and to lie beyond our control. This granted, the resulting moral law of self-control was held to be both natural and reasonable, and to apply universally, regardless of local ties and loyalties.

This Stoic universalism was in many ways an admirable view of morality. But the Stoics' attempt to take nature as the guide to principles of human conduct, combined with their circumscribed view of human freedom, was unable to inspire improved social arrangements (let alone goals such as conservation), and tended to advocate compliance with the status quo instead. Indeed this failure illustrates the severe limitations of attempting to reason from nature (or from human nature) to practical ethical principles for human conduct. (Nevertheless, Carmen Velayos Castelo and Alan Holland have each recently advanced more positive attempts to synthesize Stoicism and environmentalism.)

Certainly our nature is what facilitates reasoning in the first place. Yet sooner than reasoning from nature to morality, it is better to reason instead from the holders of moral standing (discussed later in this chapter), and their well-being and needs. But first we need to reflect on environmental problems and, to do that, on the concept of the environment itself.

The environment

Environmental problems are problems arising from human interactions with the natural world. They include pollution, depletion of resources (including fresh water and fish-stocks), degradation of land, loss of biodiversity (cultivars, wild species, and habitats), and global warming. Different understandings of such problems turn on what we consider valuable. But people also diverge in their understanding of what is meant by 'environment'.

Most obviously, an environment consists in the local surroundings (natural or otherwise) of a person or community. But many environmental problems extend across environments in this sense, and so this cannot be the only sense of 'environment'. Besides, not all such environments are worth preserving, and some, rather than falling foul of environmental problems, are the products of environmental problems themselves (think of the Dust Bowl region of the American prairies in the 1930s). Environmental concern clearly has some further focus.

By contrast, some thinkers and writers regard someone's environment as what that individual perceives as her native setting, the familiar nooks, crannies, and pathways of home territory, to which people are committed with a 'pre-ethical' commitment before we embark on any kind of ethical reflection. But not everyone has an environment in this sense, since many people lack a sense of being at home in the place where they find themselves (or in any other). Besides, environmental concern often arises for environments that we adopt rather than those we are born to. Indeed much environmental concern relates to widespread or even global problems, and thus transcends what individuals may regard as their home territory or patch.

The concept of 'environment' has also come to be applied to the objective system or systems of nature, such as mountains, valleys, islands, oceans, and continents, and the natural cycles and processes that shape and reshape them. The environment (in this sense) encompasses local environments and transcends environments in the 'home territory' sense. Indeed local environments and perceived environments could not exist but for the natural cycles and processes that make them what they are. This is the sense of 'environment' used in this book, except where contrary indications are given.

The environment (in this sense) is far from invariably hospitable. In many places it has also suffered from human exploitation or neglect, some areas having become deserts, and some seas having almost disappeared (think of the once fertile lands beside the Aral Sea in central Asia, and of the Aral Sea itself). Yet the environment still makes possible much of what is valuable in our lives, and the lives of our descendants will almost certainly depend on it. While there are other reasons for environmental protection, these are some of the central reasons why we should care for our shared surroundings, the natural environment of our planet.

If environments were invariably local, the existence of global rather than local *environmental* problems would be hard to understand. As things are, besides local environmental problems, such as flooding from local rivers, there are also global ones. Some are global because of their recurrence all over the globe; an example is traffic congestion. Others are systemic, arising from human impacts on global systems. Examples include acid precipitation, ozone depletion, and global warming.

Only through the concept of the environment as an objective natural system can we make sense of such environmental problems in the first place, as Nigel Dower has argued. Our having this concept opens the way to addressing these problems. Fortunately, ozone depletion is being tackled and, it seems, solved as a result of the internationally agreed Montreal Protocol on Substances that Deplete the Ozone Layer (1987); this offers the hope that the others can be tackled as well.

Moral standing

We can now turn to the concept of moral standing. As we have seen in Chapter 1, the question of moral standing concerns which beings matter where ethics is concerned, and should be taken into account when decisions are being taken. Much turns on the answer to these questions for our understanding of ethics in general and of environmental ethics in particular, because different ranges of affected entities will be given consideration. Goodpaster's own answer, that moral standing belongs to whatever has a good of its own and can be benefited, now needs to be developed further.

Goodpaster in fact wrote not of moral standing but of 'moral considerability', although these phrases have now become interchangeable. This was because he wanted to answer a question previously raised by Geoffrey Warnock, concerning the conditions of 'having a claim to be considered', considered, that is, 'by rational agents to whom moral principles apply'. If something should be considered, then it can be said to be 'considerable' (in a sense that has nothing to do with its size or extent). But what is more important than the language used is that such things deserve moral consideration in the sense of deserving (in Goodpaster's words) 'the most basic forms of moral respect'.

Here Goodpaster investigates whether these things are simply the holders of (moral) rights. Certainly everything that bears such rights will be morally considerable (or have moral standing). But the converse may not be true. Thus many people deny that non-human animals have moral rights, but few deny that it is wrong to treat them cruelly, or to neglect those that are subject to human care. So moral standing can be held to belong to something whose rights are in doubt, or at least not universally agreed. The notion of rights turns out to be narrower and more demanding than that of moral considerability.

The example of non-human animals is relevant in another way. Goodpaster raises the question of moral considerability partly in view of concern about the environment and its living constituents. While some might be inclined to assume that only human beings are morally considerable, environmental concern, he suggests, requires a broader view, for which the other living creatures of the planet count as well, not least non-human animals.

To vindicate this view, he considers possible criteria for being morally considerable. Being rational cannot be necessary, or human infants would be excluded, together with most non-human animals. But being sentient (having feelings) seems not to be necessary either. For many creatures are capable of well-being and of flourishing even in the absence of sentience.

The criterion that Goodpaster favours is that of having a good of one's own (that is, having interests not deriving from those of other creatures); for all such things can be benefited or harmed, depending on whether their good is advanced, preserved, or subverted. Besides, beneficence (or fostering the good of other beings) is central to morality, and so it is appropriate for all objects of beneficence to be recognized to have moral considerability. Here, Goodpaster's reasoning seems to be importantly right.

But, as was foreshadowed in Chapter 1, acceptance of this criterion means that moral considerability (or moral standing) belongs to all living creatures; and this is Goodpaster's main conclusion. Probably he had in mind the living creatures of the present. But if we bear in mind that future living creatures will equally have a good of their own, and are or will be liable to be affected by current moral agents, the implication is that future living creatures are morally considerable too.

Including this huge range of creatures among the bearers of moral considerability may appear to stretch the bounds of morality unduly, and at the same time to make the moral life impracticable. But Goodpaster anticipates this apparent problem, by making a key distinction between moral considerability and moral significance. The moral significance of a creature concerns its moral weight, and thus the degree of consideration that it deserves, relative to others. But this is a separate issue from whether it warrants moral consideration at all in the first place.

The moral significance of (say) a tree may be slight, and insufficient to outweigh that of a sentient creature (such as a bird or a squirrel). So recognizing the moral considerability of a creature does not oblige us to prioritize it. Nor does it confront us with myriads of apparently impossible choices when there are conflicts of interest between several creatures which are all bearers of moral considerability. What makes such choices possible are the different degrees of moral significance of the different creatures that we encounter or affect. The wide scope of moral considerability, however, is fully consistent with such differences of moral significance.

In other words, recognizing the moral standing of living creatures does not make moral decisions impossible, or morality impractical. Instead, it enriches our understanding of the context of action and of moral decision-making, and redirects us to include in our deliberations our impacts on living creatures of other species as well as on human beings.

But to accept the moral standing of all living creatures is to endorse a 'biocentric' stance (mentioned already in Chapter 1), or biocentrism. Biocentrism is a life-centred ethic, and holds that all individual living creatures have moral standing.

There are some biocentrists (such as Paul Taylor) who hold that all these creatures have equal moral worth. But this view conflicts with that of Goodpaster, who holds that they have differing degrees of moral significance. This position of his does not, however, discard all considerations of equality; for it is consistent with the principle of Peter Singer that equal interests should be given equal consideration. (Different creatures have different capacities and interests, but where similar interests are held, equal consideration should be given.)

Adherents of biocentrism (biocentrists) need not deny moral standing to organized groups. For example, most people accept that companies and countries have moral rights and at the same time moral responsibilities. If they did not, then they could not be held to have responsibilities to uphold environmental standards. But, this being so, they must be held at the same time to have moral standing; and biocentrists need not say otherwise.

At times, however, Goodpaster shows signs of going beyond biocentrism and recognizing moral considerability in species and in ecosystems. The issue here is whether these entities have a good of their own, and should be regarded as living beings themselves. While Goodpaster seems half-inclined to accept all this, most people are not prepared to go this far. So biocentrism is normally held to apply to those who accept the moral standing of individual living creatures, and not that of species or ecosystems. The view that species and/or ecosystems have moral standing as well as individual living creatures carries the distinct name of 'ecocentrism'.

Biocentrists and ecocentrists each have characteristic stances about where intrinsic value is to be found, as (come to that) do anthropocentrists. But before we can get to grips with such matters, or with other questions about value (intrinsic or otherwise), we need to consider the key concept of value itself.

Value

Things are valuable when there are reasons to promote, preserve, protect, or respect them. So discovering that something has value means that we have reasons for positive attitudes and actions in its regard. And when we understand something's value and have such reasons, we can go beyond issues of moral standing to issues of deciding what policies we should adopt and what forms of action we should take.

Some people suggest instead that value simply belongs to whatever is valued. But this view disregards the need for there to be reasons for valuing whatever is valuable. Indeed much that is valued has either low or negligible value, and has only been valued as a result of a passing fashion or through misplaced advertising. Further, much that is valuable is not yet valued, in many cases because it is a valuable creature that has not yet come to people's attention (or been discovered at all), or a valuable work of art that has not yet been properly exhibited or performed. Things can thus be valuable without being valued; there will be reasons for valuing them, but reasons are not always noticed or heeded.

It is sometimes objected that to stress value is to appeal to financial or economic considerations, and that forms of environmental ethics based on value must therefore be commercializing distortions. But to object in this way is to focus on just one kind of value, and ignore the others.

Money, admittedly, is useful because of its exchange value; but it lacks the kind of value that belongs to a panorama or to a sunset, or the kind that belongs to health or happiness. Its value is instrumental, but, like much else that has instrumental value, it does not of itself make life worthwhile, even though it is sometimes worshipped as if it did. And like everything that has instrumental value, what makes it valuable lies beyond it. Forms of environmental ethics based on value focus not on quantifiable kinds of value such as money, but on the value to be found in the well-being and flourishing of living creatures.

This brings us back to Rolston's distinction, noted in Chapter 1, between instrumental value and intrinsic value. The value of what is valuable instrumentally is derivative value, dependent on and deriving from the value of something other than itself. Such value contrasts with value that is non-derivative. Things with value of this latter kind are valuable because of their own nature. They may have additional kinds of value as well; thus education can be both valuable in itself and valuable because it leads to gainful employment. But their nature is such that they are valuable as ends, and not as means alone; rather than having value only as means, they give value and point to measures, means, and policies devised to attain them.

There are other kinds of derivative value alongside instrumental value. In my view, aesthetic value, as in the examples of the value of a panorama and of a sunset, is dependent on appreciation by human or other perceivers. Not everyone agrees, but the possibility of multiple kinds of derivative value should be recognized, for not all derivative value is valuable simply as a means to something else. A further kind of value that may well be derivative is symbolic value (like that of a hand-shake), which is dependent on the perceived meaning or meanings of what is performed or enacted. However, there are many things whose value is neither instrumental to nor derivative from other things we value for their own sakes. A plausible candidate is happiness.

Happiness is widely agreed to have intrinsic value. But there is more to life than happiness, and many have located intrinsic value more broadly in human well-being or flourishing. Aristotle, for example, began his *Nicomachean Ethics* with the claim that all action aims at such flourishing, a claim that he regarded as a truism. Shortly after this he made the distinction between what is desirable instrumentally and what is desirable intrinsically or in itself. There he added the reasoning that it is impossible for everything desirable to be desirable instrumentally, as there would then be nothing to give anything its desirability or its point; so something must be intrinsically desirable. And this, he argued, is human flourishing.

But most people recognize that the well-being of other creatures matters as well as that of human beings; and, once other living creatures are recognized as having moral standing, it is difficult to avoid accepting something already suggested in Chapter 1, namely that their flourishing has intrinsic value alongside that of human beings. Even if their flourishing has less moral significance, perhaps because of their different capacities, their moral standing strongly suggests that their well-being or flourishing is desirable or valuable intrinsically. For where a creature has moral standing, there must be something about it that counts as a reason for action, and a non-derivative one at that.

Some people take the view that it is sentient creatures (creatures with feelings) and they alone whose well-being or flourishing has intrinsic value. (This view is sometimes called 'sentientism'.) Organisms of this kind, it is suggested, have a conscious perspective, and what happens to them can matter from that perspective, unlike the goods and the harms that befall non-sentient ones.

But the absence of conscious perspectives from other living creatures does not make it acceptable to do them harm, something effectively recognized by the widespread response of revulsion to Routley's Last Man thought-experiment. In needlessly cutting down a healthy tree, this imaginary person would be doing needless harm and acting wrongly; such, at least, is a widely held response to this action. A likely explanation of this judgement consists in the widespread assumption that the well-being or flourishing of non-sentient creatures (such as trees) matters independently or has intrinsic value, as well as the flourishing of sentient ones.

This stance about value corresponds to the stance that recognizes moral standing in all living creatures. As has just been mentioned, where a creature has moral standing, there will be non-derivative reasons for promoting or preserving its well-being. So the biocentric stance that recognizes moral standing in living creatures also recognizes intrinsic value in their well-being or flourishing. In doing so, it diverges from sentientism, which stops short at sentient creatures and their well-being, as well as from anthropocentrism, which stops short at the flourishing of human beings.

In connection with value, one further contrasting stance should be mentioned, that of ecocentrism. This is the view that an ecosystem, as a whole, or species as a whole, has an identity and a good not reducible to the good of their members, just as a nation (e.g. Wales) or a people (e.g. the Welsh) are sometimes said to have an identity and a good distinct from that of their individual constituents. Whole forests are held to have value of this kind, and not only individual trees or their flourishing. For obvious reasons, this stance is sometimes alternatively known as 'eco-holism'.

Yet ecosystems continually fluctuate, and it is less than clear how to identify such systems. This being so, it is unclear how to understand and recognize an ecosystem's good, as both James Sterba and Emma Marris have argued. And as for species, these can be understood either as populations or as abstractions.

But abstractions can hardly be ascribed intrinsic value. As for populations, where the well-being of individual creatures is recognized to have intrinsic value, there is no need to ascribe to populations of such creatures an intrinsic value additional to that of their members. Ecosystems and species can instead be seen as matrices within which intrinsically valuable individuals emerge, and as being valuable on that basis. Nevertheless, ecocentrism continues to inspire many committed environmentalists.

We shall eventually return to these stances about value in the context of different understandings of right action. For present purposes, what is important is perhaps what they have in common, which is their positive affirmation of value and their confidence in people's ability to recognize it and to be motivated and inspired by it.

By contrast, those who deny intrinsic value, or reject all claims to identify it, are prone to make all actions and enterprises unjustified, futile, and pointless. For reasons for action are needed if actions are not to be unjustified, and it is precisely value, and ultimately intrinsic value, that supplies such reasons. Value thus supplies ethics with its grounding and its motivation, and this is equally true of environmental ethics, as all the various 'centrisms' discussed in this chapter bear witness.

Chapter 3
Future generations

The moral standing of future generations

Concern about future generations stretches as far back as the Ten Commandments, and was articulated among the ancient Romans by Cicero and Seneca, and by Dante in the Middle Ages. But the belief that present people can significantly change the future originated as recently as the Enlightenment. So does the belief that our generation may be judged by posterity, that is, by our successors.

As long ago as 1714, however, Joseph Addison asked why we should be concerned about future people, granted that they have never done anything for us. But another more recent sceptic about future-related responsibilities, Thomas H. Thompson, nevertheless grants that the questions 'Why care about future generations?' and 'Why be moral?' are in practice the same question. In any case the question 'Why care about future generations?' is at least on a par with 'Why care about those with whom we share the planet now?'.

Thus if we care about human well-being in the present, we can hardly be indifferent to that of our children and grand-children after our own life times, as if our deaths would obliterate the moral universe. In many African traditions this link is taken for granted; land belongs not to individuals but to inter-generational collectives such as clans, and any head of clan depriving coming generations of benefits they might have expected to inherit can be deposed.

When we consider the extent of current people's responsibilities to their successors, relevant future generations include all those that can foreseeably be affected by current people's actions. These extend beyond our children's generation to people of the further future, in cases where we can affect them; for the impacts of current actions are not confined to the next generation alone.

Thus if we release radioactive substances with half-lives of several centuries (or bury them insecurely), then the generations living in those centuries become relevant to our current responsibilities. As Onora O'Neill has argued, these people too turn out to have moral standing, at least from the perspective of those whose actions can affect them. But in view of the long-lasting impacts of current carbon emissions, and their often-avoidable character, the people capable of affecting these future generations turn out to be most of those currently alive.

Objections to this view: the Non-identity Problem

Perhaps the most fundamental objection to belief in the moral standing of future generations is based on the assumption that our duties are limited to making particular future people better off, and to avoiding harming them. But we simply cannot do this through changes of policy, because adopting significantly different social and economic policies (such as implementing sustainable practices) will have the effect of generating different future people. For adopting such policies means that different people will meet and have different children from those they would have had in the absence of the new policies. Hence before long no one living would have been alive if no such policy changes had been made;

and so no one living by then is better off. (Philosophers know this as 'the Non-identity Problem'.)

To express this differently, most future people cannot be harmed by current policy choices. At best, adopting new policies will generate a population that will be better off than the different population which would have lived if the policy changes had not been adopted. The individuals of the better off population would not have existed at all if the old policies and practices had persisted. So if morality simply requires us to avoid harm to particular future people, or to make such people better off, then new social policies cannot be justified on this basis. Indeed nothing at all can supposedly be owed to the future people generated by such policies.

Others, however, have challenged this assumption about the limits of our future-related duties. For we can have duties or responsibilities not only to particular individual people, but also towards whoever will be living in a certain region in a certain period of time, if we can affect the average quality of life that sets or groups of such people will enjoy. This is already assumed when we recognize duties to the people of distant places who are unknown to ourselves, duties, for example, to reduce diseases like malaria or to alleviate their poverty. Similarly if we can increase the average quality of life of groups of future people, those of us able to do this have a responsibility to do so, despite our being unable to know which individuals will lead enhanced lives (enhanced, that is, in comparison with the lives of those that might have lived instead).

Derek Parfit has argued for this view using thought-experiments. Just say we can benefit current people by depleting and consuming resources now, but all foreseeable generations will then have a much lower quality of life than if the same resources had been conserved. Given the assumption about duties being owed to particular individuals only, then the reduced quality of life of future people would be irrelevant to policy decisions about depletion, and we should forget about future people and focus entirely on the present generation. But most people who consider this imaginary scenario agree with Parfit in rejecting this implication. So the assumption on which it is based has to be discarded too. Our future-related duties are not owed to particular individuals only, but extend to whoever will live in the foreseeable future.

Accordingly the non-identifiability of most future people is no obstacle to them having moral standing. Whether or not we have duties owed to them as individuals, we can still have responsibilities in their regard.

Discounting future interests: uncertainty

Granted that future people and their interests matter, it is still widely held that their interests count for less than current interests, and that this justifies discounting future gains and losses by some fixed annual percentage (such as 5 per cent). One of the grounds often given is the uncertainty of future gains and losses. Thus cures of diseases discovered in the present may have waning effectiveness in the future, or may be outweighed by new diseases; and new flood defences may cease to be effective if sea-levels rise more than expected. Even if we recognize that future arthritis will be just as bad as arthritis today, the uncertainty about whether it will actually happen seems to lower the priority of preventive measures in the present.

But the uncertainty of some future benefits and losses does not justify discounting future benefits and losses in general or across the board. Thus many future losses are all-too-predictable, such as fatalities from malaria if it is not counteracted, and the flooding of coasts and estuaries from predictable rises in sea-level due to current rates of greenhouse gas emissions. (Laws of nature may be presumed to remain constant over time.) There again, many of the intended benefits of current policies (decisions to hold referendums, for example) are uncertain, as are many of the harms that such policies seek to prevent.

There is thus no correlation between certainty and closeness to the present. So to discount all future harms and benefits, as compared with present ones, is not warranted on the basis of uncertainty. Uncertainty warrants at best selective discounting, for cases where there are distinctive reasons to doubt current predictions and expectations.

Besides, conventional discounting involves reducing the value or disvalue of future benefits and costs year by year on a compound basis. Thus the value of the goods of thirty years hence is reduced by the agreed percentage of discounting thirty times over, and thus becomes negligible. But this approach pays no heed to the widespread concern to preserve things of value (whether works of art or natural species) for our children and grand-children. It would take more than the general uncertainty that characterizes the future to justify such a drastic disregard of what we value.

Discounting future interests: time-preference

One of the grounds given for discounting future interests is that this practice is supported by current people's preferences, as attested in empirical studies. For example, saving one life now is often considered just as worthwhile as saving no fewer than forty-five lives in one hundred years' time. Democratic decision-making, it is assumed, ought to give weight to such widely attested public time-preference.

Hilary Graham, however, finds that such results are skewed by the anonymous and dissociative language standardly used in questionnaires and surveys. If, instead, respondents are asked to compare policy options that would save lives in their own generation, their children's generation or their grand-children's generation, the results are significantly different. In response to surveys of this kind, a majority prefer options that would save an even number of lives across all three generations, and a significant number actually support options saving a greater number of lives in their grand-children's generation than in their children's, and saving a greater number in their children's than in their own generation.

Similar findings were made when the question was changed to short-term versus long-term schemes of flood-prevention. Thus a majority of respondents feel connected to future generations, conceptualized in this way, or actually identify with them. Yet the generation of the grandchildren of current adults could reasonably be held to include the period of one hundred years' time, for which studies employing anonymous terms produce quite different results.

So the evidence of empirical studies of preferences may not after all support discounting, and suggests that investments expected to benefit one's grand-children and their age-cohort are widely regarded with favour. (More empirical work would be needed to discover whether benefits to the generation of one's great-grand-children are similarly favoured.) Hence, even if we grant that public preferences are morally relevant to decisions about discounting, it is unclear that such preferences uphold this practice, as opposed to rival practices such as long-term investments to benefit future generations, and long-term policies of preservation, whether of art-works, places of natural beauty, species or their habitats.

Further grounds are sometimes given for discounting future costs and benefits, such as the possibility that future people will be better off than the present generation. But this assumption seems a good example of predictions and expectations that there are

34

ample reasons to doubt. Assumptions like this one are sometimes used in support of consuming resources in the present; future people, it is suggested, can be relied on to develop sufficiently improved technology to generate substitutes. Yet many resources turn out to be irreplaceable (think of species and ecosystems); and the only way of ensuring that resources such as rare minerals remain available is through preserving stocks of them, just as people are already preserving seeds in the Svalbard Global Seed Vault.

In general, the grounds forwarded for discounting fail to justify discounting blanketwise or in general. At best, they justify selective discounting, where specific reasons (whether of uncertainty or opportunity costs) can be shown to be distinctively relevant. This finding opens up many large issues of how to take seriously our future-related responsibilities.

Future preferences and needs

But if the interests of future people are a function of their preferences, then the unpredictability of their preferences, attitudes, and tastes means that, much as we might wish to take their interests into account, we simply cannot, because of our ignorance. Some philosophers consider this an obstacle to preserving anything for their sake, whether wild species, sites of natural beauty, or works of art such as paintings, statues, and musical compositions. For if the tastes of future people turn out not to overlap with our own, then our best efforts at transmitting a worthwhile legacy will be worthless.

Some of the same philosophers suggest that we should accordingly lay plans to ensure that future people are educated so as to appreciate, for example, the diversity and the beauty of the natural environment. This is a theme worth returning to, even if the problem it is intended to solve proves to be a misconception.

The supposed problem, however, must be just that: a misconception. For we can foresee many future interests, such as the interest of future people in shelter and clothing and in a reliable food supply, and in inheriting a relatively unpolluted environment. These were, in effect, the kinds of foresight assumed to be available to us in the course of the discussion of discounting future benefits and losses (presented earlier in this chapter). But what we can foresee is not so much preferences as human needs. To lead a decent human life, future people will require the satisfaction of the needs just mentioned, among others, even if they would not all prefer to be dependent on such needs; and it is the satisfaction of their needs (rather than of their preferences) which will be central to their interests.

Admittedly we would be unwise to impose on them our own interpretations of such generic human needs, such as particular styles of clothing or of diet, because they may have different preferences from our own. But if we provide for their generic human needs, we are unlikely to be entirely flouting their values. For they are likely to align at least some of their preferences and valuations to the needs that are common to humanity at all times and places.

Accepting the need of future people for a relatively unpolluted environment already tells us a good deal about what kind of provision we should make in their regard. For example, the Montreal Protocol (1987), which banned the use of chemicals known as CFCs (chlorofluorocarbons) and HCFCs (hydrochlorofluorocarbons), so as to preserve the ozone layer and protect people (and other creatures), present and future, from skin cancer, laid the foundations for fulfilling this need, as long as the parties to the Protocol continue to observe it. The same is true of the Kigali Agreement of 2016, which banned chemicals known as HFCs (hydrofluorocarbons), introduced as substitutes for the ones banned at Montreal, but subsequently found to be equally pernicious.

If, further, E.O. Wilson's theory of biophilia is correct, and human beings share a deep-seated need to be associated with living creatures and green spaces, then a wider range of environmental provision for future people would be in place. Already town and country planners in developed countries attempt to provide for this apparent need, and related ones such as access to open spaces for play and recreation, and for getting away from crowds; and, if the biophilia theory turns out to be correct, such provision is going to be needed not only in the contemporary North but worldwide and across the centuries.

Here it is relevant to add that the needs of future generations include the needs of future generations of non-human species, if, as was suggested in Chapters 1 and 2, these species have moral standing. The members of only some of these species will have preferences, but all will have needs both for survival and for healthy functioning, such as the preservation of the kinds of habitats and climates on which they depend. If these needs are neglected and species continue to become extinct, then billions of valuable lives that might have been lived will be irretrievably lost.

Many current species, if not eliminated in the present and near future, actually have the ability to outlive humanity, and thus to preserve the presence of life on our planet when humanity has left the scene. Neglecting the needs of non-humans thus turns out to be speceicide, or the avoidable destruction of whole species, and often of whole clusters of species. To express matters differently, the erasure of the future possibilities of these entire forms of life is, as Rolston has remarked, not just the killing of individuals, but the 'super-killing' of whole kinds, and all possible individuals of these kinds for all time.

Some future-related policies

Taking into account all these future needs alongside current ones is not going to be easy. Nor do things get easier when we bear in mind the extent of unsatisfied human needs in the current world, and the importance of putting this right. Solutions to this problem will be considered in Chapter 8 in the context of mitigating climate change, but aspects of these problems can be mentioned here to indicate how harmful practices and neglect detrimental to human and non-human health, present and future, can be avoided, and better practices introduced.

The harmful practices that I have in mind include carbon-based energy generation. Many people in the current world are obliged to heat their homes and their food using fuel-wood and other forms of biomass, and in their efforts to obtain fuel cut down woods and forests, while others remove swathes of forests to build mines, roads, dams, and smelters. (See Figure 3.) But deforestation destroys key habitats of wild creatures, and aggravates the problem of carbon emissions, while domestic fires, besides contributing to this problem, are widely a source of urban fog and of pulmonary health problems.

3. Tropical forest beside Erawan Waterfall, Kanchanaburi, Thailand: forests must be preserved for their biota—and for our grand-children.

The problem of domestic fires and the pollution associated with them can initially be tackled by the increased use of more efficient stoves, some of them burning safer fuels such as LPG (liquid petroleum gas) instead of wood and dung. Longer term solutions include the replacement of all such fuels through generating electricity by renewable processes (such as solar, tidal, wind, wave, and hydro-electric energy), which will also reduce the overall rate and extent of deforestation. Such processes can be sustained indefinitely, thus providing solutions not only for current people but also for their successors.

The disposal of toxic substances in the current world is all too often achieved by dumping them on waste-tips in developing countries, or in the poorer districts of developed ones. Such processes increase both inequality and ill-health. But they are replaceable by practices of burial in safe underground repositories, well away from vulnerable people and other creatures. Substances for which no safe form of burial is known should not be generated at all (the by-products of nuclear energy generation may be a case in point); otherwise the disposal of toxic substances can be made sustainable, in ways that avoid poisoning future generations.

Similarly soil-erosion and the growth of deserts can be tackled by sustainable processes such as judiciously selected programmes of tree-planting. Processes of this kind, once established, stand to benefit coming generations, as well as the atmosphere they will breathe. These are just some of many environmental problems and remedies, but supply examples of ways in which present policies can prevent future interests being undermined, particularly policies which, once introduced, can be sustained indefinitely.

Engaging new generations

While the justification of providing for future needs does not depend on future people's attitudes or on predicting them in the present, those people's attitudes to the world around them are still going to be crucial. For example, it will be for them to decide whether to persist with salutary sustainable processes or not; and their willingness to preserve wild species and habitats is likely to depend on whether or not they appreciate the wonder and the variety of planetary wildlife, both distant and local. Nor will commitment at government level be effective if it is not democratically grounded and widely shared among citizens.

This serves to underline the importance of widespread environmental education at all levels from students' earliest years onwards. Appreciation and love of nature are much more likely to be fostered through frequent outdoor visits to parks and wild habitats than through lessons restricted to classrooms. Yet there is also a role for wildlife programmes on television; and, as Martin Hughes-Games has recently argued, such programmes need to draw attention to human impacts on wild species, and the need to reduce them, on pain of appearing to connive at the ongoing decimation of such species.

Such education can also further policies seeking to stabilize the human population sooner rather than later. Educating girls and women (in particular) about limiting the sizes of their families is already contributing to the demographic transition from large to smaller families taking place in most continents, and the emergence of a world of zero population growth, necessary if human impacts on nature are to be curtailed, depends in large measure on the continuation and spread of such education.

If succeeding generations are to be engaged in policies of conservation, sustainable living, and care for the natural species of the planet, much clearly depends on the early, effective, and sustainable introduction of environmental education.

Representing future generations

Future generations stand to be affected by current decisions, but are unrepresented in almost all current decision-making bodies. Yet Article 1 of the 1997 UNESCO Declaration on the Responsibilities of the Present Generations towards Future Generations declares that 'The present generations have the responsibility of ensuring that the needs and interests of present and future generations are properly safeguarded.' Ways of safeguarding these needs and interests clearly need to be found.

Part of the answer consists in ensuring that future generations will inherit ongoing democratic institutions, committed to upholding social justice, human rights, and a positive quality of life, environmental quality included. If these institutions are lacking or become moribund, then future generations will have to repeat the struggle to establish them. Yet such institutions, while indispensable, are far from sufficient for future needs and interests to be satisfied.

Another part of the answer involves putting in place institutions with the role of the long-term planning of infrastructure and supplies of energy and fresh water. While there is widespread resistance to such centralized planning, private enterprise itself turns out to depend on a reliable infrastructure and reliable public services, which should as far as possible be based on renewable sources. For many countries, international collaboration will be required to establish such sustainable systems.

Yet further measures are going to be needed. For example, measures are required to conserve options for future generations, preserving not only the quality of the environment, but also cultural facilities such as theatres, museums, and libraries. Constant temptations arise to curtail such options and facilities, for the sake of short-term interests, and champions of future interests are needed to resist them. There is thus a case for proxy representation of future generations.

One proposal is that each legislature should include small numbers of members appointed to protect future interests, supported by research teams charged with researching future needs. The presence of voices speaking for coming generations could well enhance decision-making, as Kristian Skagen Ekeli has argued. But there is a danger that the remaining legislators would leave representing the future to those appointed to do so, and pursue short-termism instead.

A further problem concerns how to appoint representatives of an electorate that does not yet exist. This might be overcome by an agreement to appoint these representatives from future-oriented pressure groups, and also from preservationist pressure groups (so that the interests of future non-human creatures are heeded as well), provided that these groups themselves satisfy democratic criteria; yet these representatives would still remain open to challenge by legislators appointed in democratic elections among current voters.

Another possibility would be the appointment of an ombudsman as a watchdog of future interests. A good example is the Hungarian parliamentary commissioner, who is entrusted with monitoring legislation 'to ensure the protection of the fundamental right to healthy environment', with conducting 'investigations into potential or alleged violations or threats to the environment and future generations', and reviewing the actions of municipal and local governments as well as the national government, and empowered to halt or modify governmental actions accordingly. While such an office could generate controversy, and might be open to abuse, its challenge to vested interests could well be salutary.

Quite a different possibility would be the legislative device of granting legal rights to ecological systems (as once suggested by Christopher Stone) such as rivers. During 2017, rights were granted to the rivers Ganges and Yamuna in India and to the river Whanganui in New Zealand. While there are dangers here, one of which is that pollution control could be made more difficult, there are also clear benefits to the future human users of these rivers and also to the river creatures of the future. Without attributing moral standing to entities such as rivers, biocentrists can applaud such protections for the sake of future people and other creatures that stand to benefit.

One final possibility is the enshrining of future-related guarantees within a country's constitution. Thus, as the World Future Council reports, the Ecuadorean Constitution guarantees 'a sustainable model of environmentally balanced development...to conserve biodiversity and the natural regeneration capacity of ecosystems, and to ensure the satisfaction of the needs of present and future generations'. Similarly the South African Constitution affirms the right of everyone 'to have the environment protected, for the benefit of present and future generations'. These provisions have not prevented controversies about their interpretation and application, but they enshrine a dependable way of concentrating minds on issues that might otherwise be neglected.

Chapter 4
Principles for right action

Moral knowledge

Environmental ethicists, like those in other branches of ethics, cannot escape from considering what ought or ought not to be done, and how this is to be decided or discovered. Fortunately, we are not confronting these issues here from scratch, having already reached conclusions in Chapter 3 about obligations to future generations. These findings may now help us reflect on how to understand moral principles. For example, accounts that uphold obligations to future generations should be preferred to ones that have scant regard for such obligations.

An obstacle to making progress with moral principles is the widespread belief that issues of what ought or ought not to be done are all matters of opinion, and that they do not admit of knowledge. Much ink has been spilt in discussions of these matters, and these discussions must here be set aside. Suffice it to say that most people recognize that there is such a thing as knowing the difference between right and wrong, and that, as such, it must sometimes be possible for moral claims to be true or correct. It could still be that many such claims are too vague or overgeneralized to be reliable, and that great care is needed before we can claim to have attained moral knowledge. Yet the very possibility of moral knowledge should encourage us to continue to look for it, rather than to despair of the attempt.

Some find a problem in the vagueness of the word 'ought'. This impression may well be due to 'ought' being used on different occasions as short for several different kinds of 'ought', such as 'prudentially ought' (ought in one's own interests), 'legally ought', 'technically ought', 'aesthetically ought', and 'morally ought'. 'Morally ought' will mean something like 'Ought in the interests of all the parties with moral standing that are affected', and this is obviously different from 'ought if you are to obey the law', 'ought if you want to use the best technical means to your ends', and so on. Thus while unqualified 'oughts' may be 'oughts' of any sort, and can be disagreed about accordingly, moral 'oughts' are much clearer. Seen in this light, moral knowledge becomes possible.

Some people feel that they can sometimes be sure about particular judgements (like what they should do for their young children), but not about principles, which hinge on words like 'all' or 'none'. It is true that most principles have inbuilt exception-clauses, such as 'except when this principle clashes with an equally basic principle', or 'except in highly exceptional circumstances'. But principles such as 'Promises ought (all) to be kept' remain reliable for the generality of cases, and can be known to be so. (Why this is so would bring in philosophical theories, but these are generally less reliable than principles like this one.) Hence there is no need to despair about the quest for moral principles, despite their generality, even though they include words like 'ought'.

The contract model of ethics

Some philosophers, including John Rawls, have suggested that principles and judgements are acceptable and fair which would be agreed by rational and self-interested individuals, knowledgeable about human life in general, but ignorant of their own life-prospects. This thought-experiment was devised to avoid possible bias from (say) social privilege or class interests. So we imagine that when we sit down to decide what social arrangements are fair, we go behind a 'veil of ignorance' that causes us to forget whether we are rich, poor, able-bodied or disabled, young or old. We imagine further that we do not know which society we are going to live in, let alone which family or gender, but that we just know we will need to live together in the same society and generation.

This, then, is a contract theory, based on what would be agreed in the conditions just described. In conditions of such uncertainty about our futures, Rawls argues, our only rational recourse is to vote for principles and policies that provide equal protection for everyone's rights, ensure fair opportunity to lead the lives we prefer, and promote improvement in the welfare of the worst off (for we might be among them).

This approach can have merits when groups of people need to devise shared rules to govern their relations. But it performs less well when future generations have to be considered (as it has been argued in Chapter 3 that they must). To provide for future generations, Rawls initially modified his assumption that the contracting parties are self-interested, and makes them concerned for their own descendants or 'lineage', as well as for themselves as individuals. With this motivation, they would, he suggests, select a Just Savings Principle, such that each generation invests to benefit its successor, and that succeeding generations will be no worse off than their predecessors. In *Political Liberalism* (1993), however, Rawls replaces this principle with the selection of whichever principles of resource conservation and distribution (potentially affecting future generations) members of any generation would adopt as principles they would wish both their own generation (and succeeding ones) to follow, and also previous generations to have followed.

Here we might want to add several other principles to this rather threadbare account of future-related obligations, such as preserving a decent environment and banning time-bombs. But the real problem lies elsewhere. For (apart from the way he attempted at one stage to modify people's motivations so that they would care for their 'lineage', an attempt that he later discarded) his choosers are presented as having no social bonds, such as family ties or friendships. They remain what the feminist philosopher Seyla Benhabib has called 'disembedded and disembodied individuals'. As she adds, there is no reason to suppose that what such individuals choose would be just or right, with regard to their own or subsequent generations, even if their motivation is modified to care for their 'lineage'. Nor, we might add, can there be fairness or justice if social arrangements are based on Rawls's conception of everything other than humans being mere resources to be 'fairly' distributed.

Some have suggested that the problem lies in making the choosers members of the same generation, and it may be alleviated if they are instead representatives of all coming generations. But this does not solve Benhabib's problem, and it also falls foul of another. For this suggested improvement assumes that it is already known how many generations there will be. Yet this information itself depends in part on which rules the choosers select for their society to live by. So the suggested improvement turns out to suffer from circularity; the choosers have to decide something that has to be settled before they can start.

Other contract theories have been put forward. But they all face a problem already hinted at, that of providing fairly for non-human creatures. For they all turn on contracts being made by rational, communicative, language-using persons, thus excluding from participation creatures of other species. If these creatures are provided for, it will only be through persons requiring this; but this will require not only a degree of altruism that contract theories are reluctant to assume before social rules have been formed, but also a degree of insight and understanding of the needs of other species, not readily attained and certainly not to be relied on. In other words, contract theories all break down over inter-species equity.

Contract theories can still serve as a useful model for international agreements, because if people can agree to international rules whichever society they may find themselves belonging to, that is some indication that the rules thus chosen are fair. Some theorists have modified Rawls's contract in this direction, and applied the modified theory to issues like international trade and to sharing the waters of international rivers. But contract theories continue to fail to ensure equity between generations and between species.

Virtue ethics

A more promising approach is that of virtue ethics. Character, its adherents suggest, is more important than right action, because it makes people dependable and trustworthy, and more likely to behave consistently and fairly in future than adherence to moral rules. Many of its supporters derive their account of virtues from Aristotle, who (with great insight) represented virtues (and vices) as stable dispositions resulting from sequences of choices, and as involving practical wisdom too. Accordingly, for Aristotle, the virtues are those traits of character essential for our becoming the best, most well-rounded and fully developed persons we can be, and resisting impulsive passions (fear, avarice, etc.) that undermine our capacity to conduct our lives wisely. The virtuous person is likely to be one who has been well brought up. Right action, for this approach, is simply the action that the virtuous and well brought up person would adopt.

There is much to be said for this approach. If we ask ourselves whether we are behaving as a courageous, kind, humble, and fair-minded person would, then our deeds are unlikely to be knowingly damaging, let alone disastrous. Besides, as Rosalind Hursthouse has argued, the virtues can be interpreted as taking into account both future people and animals, because these can be objects of kindness and fairness. The theory is prone to hold that it is our intentions (or perhaps our motives) that matter; but these are in any case widely regarded as central to acting as we should.

But this approach appears weaker where the unintended consequences of actions are concerned. Many of these are foreseeable, and we can plausibly be held responsible for foreseeable consequences, whether they were intended or not. Thus actions that repay international debts at the cost of depriving future generations of forests destroyed to fund the repayments may well be misguided, despite appearing virtuous; and the same applies to many other ways of deploying resources that disregard the needs of future people and/or other species. Another example is buying a diesel-powered car on environmentalist grounds because of low carbon emissions, without regard to its emission of particulates and nitrous oxide, thus contributing to dangerous levels of air pollution.

The virtue ethicist might reply that lack of care for the future unintended consequences of our acts and omissions is itself vicious. But if so, what is being offered is an unusually demanding account of the virtues and the vices, absent from accounts of the virtues and vices for example in Aristotle, who (as was mentioned in Chapter 1) held that we have to disregard the impacts of current actions on future generations, on pain of making ethics too complicated and difficult. In any case this kind of account of the virtues and vices goes beyond basing morality either on intentions or on motives, and effectively relies on the moral significance of the future consequences of present actions.

This already shows that there must be more to morality than intentions or motives. There again, as we have seen, only some interpretations of the virtues involve taking future generations and other species into account. And in times of rapid technological change, behaving as the well brought up person would do may involve catastrophe, through lack of regard for serious and/or irreversible impacts. So it does not look as if virtuous behaviour as such is bound to be right or morally justified.

Besides, as even champions of virtue ethics, such as Hursthouse, allow, no human being is perfectly virtuous. All of us are prone to moments of weakness of will when we are unable to live up to our own standards, or depression when we cannot rise to challenges besetting us, or confusion in strange or emergency situations. Hursthouse suggests this is one of the points of common sense moral rules. Such rules serve as reminders of how a virtuous person will conduct herself even when her virtuous dispositions temporarily fail her, and thus at least avoid outright virtuous conduct she would later regret. Another point of such rules is their value for accustoming the young to what virtuous conduct is like before their own characters are fully developed. So even for virtue ethics, well-chosen, justifiable moral rules are essential.

But this is an important concession, because the justification of these moral rules cannot itself turn on virtues, but must turn on something beyond them. Rolston has recently advanced a parallel response: it is values (including the intrinsic value of nature) that give the virtues their point, and not vice versa.

Yet, as Dale Jamieson has argued, virtues whose overall pursuit benefits people and other creatures of the present and future are worth cultivating, and may form part of an acceptable account of right action. Certainly outcomes are likely to be better if people stick to beneficial traits of this kind rather than calculating the consequences whenever a decision has to be made; so great is the risk of miscalculating them. But right action should also take into account both moral practices, the benefits of which turn on their widespread acceptance and thus their reliability, and also our ability to reflect on the foreseeable effects of policies (both individual and collective). And all this takes us beyond virtue ethics and back to the place of moral rules.

Rules and duties

Several philosophers have proposed that rightness means compliance with certain rules, which are either self-evident or required by considerations of reason and consistency. The kind of rules they have had in mind are socially pivotal rules about refraining from murder, keeping promises, and not telling lies. (Sir David Ross once produced a short-list of five rules of this kind.) This approach is often contrasted with one that appeals to the consequences of rules or actions to justify them ('consequentialism'). Because of its stress on duties (and sometimes on performing them just because they are duties) it is known as 'deontology' (from the Greek word for 'duty').

Sometimes the favoured duties are treated as self-evident, or as known more securely than either would-be justifications or would-be criticisms. As such, they apply to people whether they have virtuous dispositions or not, and this claim appears to count in favour of the approach we are considering. Yet those who favour this approach ('deontologists') still need to say what people should do when the rules clash, and also whether they admit of exceptions, as when a catastrophe could be prevented, but only through breaking a promise.

They also need to be able to explain how new rules can be introduced to deal with new kinds of issue, such as conservation and recycling, for otherwise their approach will fail to respond to new problems, including environmental ones. But all this suggests that it may be necessary to be able to appeal beyond the rules, if they are to be justifiable and reliable.

Immanuel Kant suggested that we let reason be our guide. It is irrational to adopt rules that could defeat the objectives for which they were adopted; so, rationally speaking, we should categorically reject any such rule. (This he calls the 'Categorical Imperative' of morality.) Say we are tempted to endorse a rule allowing us to force others to do as we please, without their consent, and thus to treat them as mere 'things', whenever it served our personal objectives. Before we do, Kant held, we must ask ourselves if adopting such a rule could prove self-defeating. Moral rules are meant to be universally applicable, and so we must ask whether the rule could prove self-defeating if universally adopted. Clearly it could. Instead of promoting personal objectives, the rule would do the reverse by licensing the overriding of personal objectives. Since it is self-defeating, reason demands we reject it categorically, whatever the consequences.

We ought instead always to respect persons and to treat them as 'ends in themselves', not mere things, and no less entitled to pursue their personal objectives than we ourselves. On the face of it, this seems like a salutary approach to choosing moral rules and principles (condemning the exploitation of human beings as it does). But its complete disregard for the well-being of non-human animals and other creatures makes it fall short of a comprehensive guide to right conduct. Besides, it is difficult to derive from it any clear guidance about obligations to future generations.

Followers of Kant have sometimes attempted to apply his teachings to inter-societal relations, which would have to be conducted, in the light of the requirement to respect persons, without force or fraud. But we can reply that his explicit disregard for the consequences of actions, rules, and practices, and the problems for his approach that this brings, illustrated in the problem-cases mentioned already, suggests that a different approach to moral rules is needed. Besides, Kant's demand that we follow rules that are not self-defeating, however dire and extreme the consequences may be, stretches credibility.

Beneficial practices, traits, and actions

Perhaps what really makes adherence to social rules and practices right (we could say, continuing our reply to Kant) is their overall beneficial character when observed with solidarity or near-solidarity. This may account for those examples where Kant's Categorical Imperative approach fares best, as with his rejection of self-defeating practices like lying, and with his advocacy of respect for individuals governing their own lives. It could also account for the rightness of adherence to beneficial practices not yet universally observed but with a fair prospect of being introduced, such as abstention from bribery, and, there again, such as recycling.

Similarly what makes character traits virtuous may well be the beneficial character of their widespread adoption. This thought echoes Dale Jamieson's advocacy of cultivating traits that benefit present and future humans, and other creatures too. It also echoes what was said about responsibilities to future generations in Chapter 3.

But we cannot extrapolate from these findings to the view that what makes actions right is simply their positive impacts. One problem is that we often cannot foresee those impacts. What then of the suggestion that what makes actions right is a favourable balance of good over bad *foreseeable* impacts? In such cases, the favourable foreseeable impacts form reasons for performing those actions, and while the unfavourable impacts are reasons against, the former are stronger.

However, the risks of miscalculation and of distorted views of what the impacts will be suggests that where beneficial practices are applicable, it is better to adhere to those practices rather than to pursue apparently beneficial infringements of them; and also that except in exceptional circumstances it is better to adhere to behaving as the virtuous person would (for example, being kind and not cruel) than to pursue the apparent benefits of flouting virtuous behaviour. But where there are no relevant beneficial practices and the path of virtue is unclear, it would be the balance of positive over negative foreseeable consequences that make an action right.

This can be called a 'consequentialist' approach to rightness, because consequentialist approaches make rightness turn on the differential impacts or consequences made by actions, traits, or practices. One well-known form of consequentialism is utilitarianism, often criticized for adopting an instrumental approach, whether to human beings or to non-human creatures. That, however, is an unfair criticism, because utilitarians seek to maximize happiness, and often include maximizing the happiness of sentient non-human creatures as well as of human beings.

But there is a big difference between utilitarianism and the kind of consequentialism supported here. For utilitarians, the sole value is happiness, and the sole disvalue is unhappiness. This stance omits all the other dimensions of the well-being of humans and sentient non-humans, and neglects the well-being of non-sentient creatures altogether. A large part of the point of the discussion of value in Chapter 2 was that intrinsic value can be found much more widely than just in happiness, and that those who care about value will take into account the well-being of humans and non-human creatures (both present and future). So even if you are against utilitarianism, you should not reject broader forms of consequentialism like this one.

Different criticisms of consequentialism concern predictability, intentions, and justice. The first of these criticisms says that we cannot make future impacts crucial in ethics because we cannot predict the consequences of actions distant in time. But we can predict the probable consequences (of actions, traits, and practices), and foreseeable patterns of impacts, in the light of experience. It is because we can do this that we can take into account the interests of future generations, as argued in Chapter 3.

Another criticism is that consequentialism plays down the role of intentions in ethics, whereas the difference between deliberate and inadvertent actions is ethically important. This is certainly an important difference, but it is important in matters of character-assessment, praise, and blame, and not in matters of rightness. For it is possible to do the wrong thing inadvertently as well as deliberately, just as it is possible to do what is right with questionable intentions. So consequentialists can recognize the importance of intentions, without making rightness or wrongness depend on them. Besides, if it is obligatory to do what it would be wrong not to do, then a consequentialist understanding of obligation (or duty) is needed too.

Finally, we should consider the apparent problem that consequentialism seeks to optimize consequences, but not to distribute benefits justly. This problem needs a fuller treatment than can be offered here. But this much can be said. Justice appears to turn on the satisfaction of basic needs, and there is nothing to stop consequentialists giving priority among values to needs that are basic over those that are less basic, and to needs over mere preferences. Hence consequentialists give due importance to justice and fairness, through seeking out policies and procedures that prioritize needs, and basic needs in particular.

I am not suggesting that environmental ethicists have to be consequentialists. Good work in environmental ethics can be done by adherents of other views (contractarians, virtue ethicists, and deontologists), just as it can by anthopocentrists as well as by biocentrists and ecocentrists. Some of the best work is done by pragmatists, who sometimes resist holding theories about rightness altogether. But where the question concerns which approach is the more consistent, most fruitful, and best serves the needs of future generations, the answer seems to rest with consequentialism, allied to a broad theory of value (as I have proposed both here and elsewhere).

Which theory of value?

Some approaches count human well-being and nothing else as valuable. An example is the Rio Declaration on Sustainable Development of 1992, which made human interests central to sustainable development, as the parties to the Rio conference could not agree on any other approach. (The Brundtland Report of 1987, which led to the Rio Conference, appears in places to recognize intrinsic value in non-human species and their well-being, but this recognition is absent from the Rio Declaration.) However, this Declaration is an example of a worthwhile contribution to the addressing of worldwide problems of sustainability, resources, population, and even of biodiversity preservation.

Nonetheless, approaches that are unable to take non-human interests into account suffer from serious limitations. An example is the limitations of contract theories, discussed in the second section of this chapter. Such theories leave out the perspectives of sentient creatures, as well as the interests of other species. But this means that they are prone to neglect problems such as whaling, factory-farming, the ivory trade, and the acidification of the oceans that is endangering coral reefs. It is true that these activities and processes can all be criticized on the basis of human interests, broadly interpreted; but anthropocentric approaches are unlikely to take these problems with sufficient seriousness. And what this suggests is that environmental problems are best addressed with a broader value-theory than contract theories or other anthropocentric theories can allow.

To recap stances mentioned in Chapter 2, broader theories include sentientism, biocentrism, and ecocentrism. Any of these could be harnessed to consequentialism, so that a broader range of values and disvalues would be taken into account.

Thus Peter Singer combines consequentialism with a sentientist value-theory, which takes into account the well-being of humans and other sentient beings, but is concerned with further living creatures only to the extent that they provide sentient creatures with habitats, livelihoods, or other benefits. His work has importantly drawn attention to the horrors of factory-farming. But such an approach has a limited capacity to protect systems of living creatures such as coral reefs. Admittedly the fish and crustaceans that live there are probably sentient, and their interests are at stake, as are human interests in the aesthetic appreciation of these environments. Yet it is hard to believe that these are the only interests at stake.

In the same way, it is difficult to credit that when a woodland is threatened, the only beings with moral standing are the animals and birds that live there, and not the trees, the fungi, and the invertebrates. If these too have moral standing, then a broader value-theory than sentientism is needed.

This brings us to biocentrism and ecocentrism. The main difference is that ecocentrists attribute moral standing not only to living creatures but to ecosystems and species as well (or, in some versions, instead). But an ethical system based on the good of ecosystems or the whole biosphere (as was proposed by Aldo Leopold), and not of individuals, cannot uphold those very rules (like keeping promises and refraining from murder) that deontologists like Ross have plausibly presented as ethically pivotal. Thus a plausible ethic should recognize (at least) the moral standing of living individuals.

But perhaps such an ethic should be concerned about ecosystems as well as individuals. Certainly the good of many individuals depends on the intactness of ecosystems in which they can flourish. But if the good of ecosystems is to be counted in addition to that of individuals, three problems arise. One is that because ecosystems largely consist of living individuals, these individuals are going to be counted twice over.

Another is that ecosystems, which are constantly changing, have no clear identity, and so it is unclear what their good consists in. The third is that many ecosystems have been modified by humans (into pasturage, gardens, and parks, for example), and that it is implausible that the good of these should count alongside the well-being of individual creatures. While such systems harbour values, the values turn on the well-being of the creatures within them, and also on that of their human carers.

Ecocentrists sometimes suggest including species alongside individual living creatures. But when considered as abstractions, there is no case for including species. The case for including them rests on considering them as populations of individuals, both present and future. But as long as the good of future living creatures is considered as well as that of current ones, then everything needing to be included has been taken into account. For example, we can now see why great importance is attached to preserving the few surviving members of a rare species, since the very existence of all future members of that species depends on them (to recap something said in Chapter 1).

This suggests that a biocentric ethic, which recognizes the moral standing of all living creatures, present and future, and the intrinsic value of their well-being, should be preferred. But systems of biocentric ethics are more plausible if they uphold actions, traits, and practices that foster value, biocentrically interpreted. Here, Peter Singer's claims that equal interests count equally, with greater interests (such as interests grounded in ampler capacities) counting for more, becomes important. This claim counts against treating the good of all creatures equally, as Paul Taylor has suggested, and recognizes that some species have greater interests than others. (It is not only, and not always, human beings that have greater interests than all others; but the greater interests of (say) primates such as orang-utans than those of insects should be recognized rather than denied.)

Taylor's approach plumps for rules not grounded in the impacts of actions or inaction, rules grounded in widespread intuitions. He extends basic human principles of justice to inter-species conflicts, requiring us to minimize or rectify unjustified injuries to other creatures. But there is more to inter-species ethics than simply avoiding outright injustice. A more cogent form of biocentric ethic bases the actions, the rules, the traits, and the practices that it favours on their foreseeable consequences. In other words, it advocates biocentric consequentialism.

Callicott's triangle, and some verdicts

J. Baird Callicott suggests that an environmental ethic contrasts with traditional (anthropocentric) humanism and also with (sentientist) animal welfarism; these stances comprise an equilateral triangle of mutually opposing positions. But both anthropocentrists and sentientists can and do contribute worthily to environmental ethics. Besides, far more than three stances are possible; as we have seen, there are different kinds of ecocentrism and of biocentrism. So we should move beyond seeing the debate as triangular. Environmental ethics is rather a dialogue between many stances and voices, and certainly not a single stance.

Others suggest adopting pluralist views, which combine two or more of the stances mentioned here. Combining several values (such as cultural preservation and nature preservation) is a salutary approach. But attempts to combine conflicting stances are liable to produce contradictions.

Others again suggest setting aside theories of rightness, in favour of a pragmatic focus on issues. Focusing on the facts is admirable, and often the theories agree over solutions (as, perhaps, over addressing climate change). But attempts to ignore theory altogether risk reaching distorted solutions, often through focusing on human interests alone.

Adherents of different theoretical stances can collaborate over practical campaigns, and about which traits and which practices are virtuous ones. But they will be more clear-headed if they have a developed stance (such as that commended here), and can defend and justify it. Such people know what they are campaigning for, and why. Sometimes they even know what ought to be done.

Chapter 5
Sustainability and preservation

Sustainable development

The sustainability of a practice or society means its capacity to be practised or maintained indefinitely, and the main point of the early advocacy of sustainable forms of society (on the part of Herman Daly and others) was the importance of recognizing limits to certain forms of growth, including ecological limits. These forms of growth included growth of production and of population. Besides, sustainable practices have the capacity to provide for the needs of future generations as well as the present. Sustainable forestry, for example, limits the annual harvest from forests to allow sufficient regeneration so that similar harvests can be taken every year into the indefinite future. Thus early environmentalists (including Daly's fellow-authors in his *Toward A Steady-State Economy* (1973) such as Kenneth E. Boulding and Nicholas Georgescu-Roegen) tended to advocate sustainability.

We should not assume that whatever is sustainable is good, much less that people who call something sustainable are always favourably disposed towards it. Bad or questionable practices can be sustainable, such as slavery and prostitution. But practices that produce overall benefits, generation after generation, such as sustainable forestry and sustainable fisheries, and observe whatever limits this requires, can be (and were) widely welcomed by environmentalists, including those who were urging humanity to live within its means. In these cases, there were also grounds against treating forests or fish-stocks simply as resources, and for leaving significant regions unharvested, and thus for restraints even on practices of a sustainable kind. Yet the sustainability of a practice soon came to be seen as a key virtue, and practices of this kind as contributions to a sustainable world society.

Development was another matter, with its customary overtones of encroachment on nature and commercialism. But given the importance of development in the distinctive sense of moving away from poverty, hunger, disease, and kindred evils, and enhancing well-being, it was recognized in the Brundtland Report of the United Nations (UN) sponsored World Commission on Environment and Development (1987), chaired by the then prime minister of Norway, Gro Harlem Brundtland, that sustainability needs to be blended with development in what the authors called 'sustainable development'. A society is developing if it is overcoming these various evils and raising quality of life in a participatory manner. Meanwhile sustainable development was defined by Brundtland as development that 'meets the needs of the present without compromising the ability of future generations to meet their own needs'. Sustainable development, understood as in the Brundtland Report, was given worldwide endorsement at the international UN Earth Summit held at Rio in 1992.

The Brundtland Report, it should be added, had more in mind than the definition just given might suggest. It envisaged social and ecological as well as economic needs, and favoured not just leaving future generations with options for satisfying their needs but also introducing policies that would make the meeting of those needs (and thus people's human rights) more feasible. Thus it favoured the introduction of sustainable agriculture and fisheries, sustainable energy generation, and, importantly, the gradual stabilization of population levels; and also provision for the preservation of species and ecosystems. Besides, it presented this case on a non-anthropocentric as well as an anthropocentric basis. However, the Rio Declaration that endorsed its proposals (the central statement arising from the Rio summit) unashamedly put human interests at the centre of its concerns.

As a result of the adoption of sustainable development at the Rio conference, both countries and companies became prone to interpret this concept in ways that suited their interests, and this led to criticisms that it had come to mean, effectively, business as usual. But this view failed to take into account the ethically radical nature of the Brundtland Report and of the Rio Declaration, accompanied as it was by the Framework Convention on Climate Change and the Convention on Biological Diversity (which enshrined the intrinsic value of biological diversity in its Preamble), and of their attempts to find an ethical pathway for humanity to follow, in matters of both development and environmental sustainability. As successive conferences reviewed these matters, the radical character of sustainable development was soon to be captured in the sets of UN goals agreed in 2000 and 2015.

Meanwhile sustainable development was criticized by Wilfred Beckerman. Construed as setting no limits to substituting natural entities (such as trees and ores) with artificial ones (like buildings) (Beckerman calls thus 'weak sustainability') it does not diverge from conventional economics, and is vacuous. But if instead it sets firm limits to substitution, for example requiring us to preserve all natural wild species including beetles (he calls this 'strong sustainability'), then, he claimed, it is morally repugnant, as the resources spent on preservation could have been spent on the relief of human poverty. In reply, supporters of sustainable development including Daly maintained that it does support preservation and thus limits substitution, without seeking to preserve every single species. The goals of poverty reduction and species preservation need to be jointly honoured (an ethically defensible approach), and where possible pursued together (as in forms of ecotourism which provide livelihoods for people of a biodiverse area at the same time as promoting preservation).

Initiatives marking the millennium

Workable solutions to ecological problems are widely held to require sustainable practices to be embedded in their planning, and to need to embody sustainable development if injustices to future generations are to be avoided. Sustainable food-production and sustainable irrigation are examples of such solutions. In their absence, future generations would be entitled to complain of neglect by their predecessors.

But alongside sustainable systems, steps are needed to overcome problems such as poverty and such as deforestation. In some cases these steps are preconditions of sustainable systems being introduced. Recognition of this prompted the initiation of the UN Millennium Ecosystem Assessment (2001–5), and also international endorsement in 2000 of eight Millennium Development Goals (MDGs), to be achieved by 2015. Of the eight goals, the first required the halving by 2015 of the proportion of people living on less than US$1.25 a day, and of the proportion of people suffering from hunger. We should recognize an ethical imperative to include these as targets, together with the goals to achieve universal primary education, to promote gender equality, to improve maternal health, to reduce child mortality rates, and to combat diseases such as HIV/AIDS and malaria. Enhancing well-being required nothing less.

The remaining goals were to 'ensure environmental sustainability' and to 'develop a global partnership for development'. But the latter goal (8) was silent on assistance by developed countries with the ecological problems of developing countries, and the former

(Goal 7), despite encouraging the integration of principles of sustainable development into countries' policies and programmes, appeared modest and less than comprehensive in its selection of targets. These targets included halving the proportion of people lacking sustainable access to safe drinking water and basic sanitation, and significantly improving the lives of at least a hundred million slum-dwellers. Yet access to drinking water and sanitation, and making slums less unhygienic, contribute importantly to the environment of the people affected. (The target for slum-dwellers was actually attained.)

The remaining target under Goal 7 was to reduce biodiversity loss by reducing the rate of loss of land covered by forest, the rate of CO_2 emissions (discussed in the final chapter of this book), and consumption of ozone-depleting chemicals. The proportion of fish-stocks conserved within safe biological limits was to be enhanced, at the same time as reductions of the proportion of total fresh water resources used, and of species threatened with extinction, and an increase of terrestrial and marine areas protected.

These were salutary aims, without being sufficient to prevent biodiversity loss even to the extent that they were achieved. (The number of individual nonhuman animals has reduced over recent decades by up to 50 per cent.) Yet the limitation of ozone-depleting chemicals has been attained, preventing vast increases of skin cancer, with the Kigali Agreement of 2016 now supplementing the earlier Montreal Protocol of 1987. And if the limitation of carbon dioxide (CO_2) emissions had been attained, that would have greatly enhanced the environmental sustainability of the planet, reducing biodiversity loss as well.

The MDGs have widely been held to underemphasize the participation of those most affected (despite the stress on participation in the UN Declaration of the Right to Development of 1986). For example, many of the world's poorest people are farmers, but agriculture was not specifically mentioned in the MDGs, let alone the issue of food sovereignty (control by countries of 'the whole food chain') for which indigenous communities and the transnational farmers' organization Via Campesina campaign. Meanwhile the Millennium Ecosystem Assessment appraised the consequences of ecosystem change for human well-being, employing an anthropocentric basis and an 'ecosystem services' approach, while affirming the need for food security, albeit not food sovereignty.

The introduction of universally recognized targets seems to have focused attention on health (including environmental health) and related issues, and to have increased the commitment of developed countries to poverty reduction. The MDG of halving the population living on less than US$1.25 per day has been attained, but, despite significant progress, the goals for child and maternal mortality, for sanitation, for education, and for halting deforestation have not been met. Meanwhile the global record on food security remains patchy and variable.

The Sustainable Development Goals

In view of the uneven progress in attaining the MDGs, and of criticisms of their content, Colombia suggested in 2011 that they be succeeded by Sustainable Development Goals (SDGs), and the UN secretary-general (Ban Ki-Moon) established in 2012 a task force to establish global goals for the period after the expiry of the MDGs in 2015. These goals were to embody all three of the dimensions of sustainable development, the environmental, economic, and social dimensions, together with their inter-linkages. This process resulted in agreement in 2015 to adopt the current SDGs.

The title of the agenda adopted by the UN General Assembly was 'Transforming our World: The 2030 Agenda for Sustainable Development'. The selected goals attempted to tackle not only global problems but also their causes. Thus the opening goal was the abolition of poverty worldwide, to be tackled in part through reducing gender inequality (recognized to perpetuate poverty). The second goal was the ending of hunger, through attention to improvements to agriculture and nutrition.

Several of the seventeen goals concern environmental sustainability. Thus the goal about health includes a target to reduce deaths and illnesses from pollution-related diseases, which include pulmonary diseases from dust-storms and smog, as well as from carbon emissions. The goal of clean water and sanitation for all could improve the environment for millions, and is widely claimed to be indispensable if any of the other goals are to be achieved, but would require extensive international funding. The related goal of making cities and communities sustainable may promote both cleaner air, urban gardening, and an increased recycling of waste products. Similarly the goal of 'responsible consumption and production' requires efforts to make both consumption and production sustainable.

Other goals relate yet more closely to environmental concerns. The goal of climate action calls for efforts to combat climate change both through controlling emissions and through promoting renewable energy (see Figure 4), thus linking with the goal of affordable and clean energy, which requires access for all to energy that is 'affordable, reliable, sustainable, and modern'. (Nothing is said about whether this is meant to include or exclude nuclear energy. But it is again clear that massive international funding will be required to introduce renewable energy systems in developing countries.)

The goal of 'life below water' seeks to 'conserve and sustainably use the oceans, seas and marine resources for sustainable development', leaving it open whether this is aimed purely at human interests or at those of marine creatures as well, although mention is made at one point of 'ocean health'. The goal of 'life on

4. San Gorgonio Pass Wind Farm, Palm Springs, California: a key
source of renewable energy—making the desert bloom.

land', however, more helpfully seeks to 'protect, restore and
promote sustainable use of terrestrial ecosystems, sustainably
manage forests, combat desertification and halt biodiversity loss'.
For these sub-goals require the preservation of ecosystems and
species, whether in the human interest, for the sake of non-human
creatures, or, as seems the likely intention, for both.

Some critics have suggested that the 169 targets of the SDGs
make them unwieldy and unmemorable. But this problem could
be overcome through suitable presentation of the seventeen
central goals. Another criticism is that these goals jointly
require growth in global production, and that this would
undermine their ecological objectives. It is true that these
goals would involve increased electricity generation, to satisfy
currently unmet human needs. Yet increases in renewable
energy generation need not subvert either species or ecosystems,
and if adopted in place of mining and excavation could help to
keep carbon-based fuels in the ground. While the attainment of

some goals could conflict with that of others, the risk of (say) all the ecological targets being undermined by success with the other targets appears slender.

Certainly, practices can only be held to be sustainable if they do not undermine other potentially sustainable practices. If, for example, would-be sustainable agriculture were to undermine significantly the habitats of wildlife and thus its preservation into the future, that would show it not to be sustainable after all. Agricultural policy makers need to bear this problem in mind, but it cannot be assumed in advance that they will fail.

An important principle for all parties to bear in mind is the Precautionary Principle, which advocates action to prevent outcomes from which there is reason to expect serious or irreversible harms, even in advance of scientific consensus being reached. (Waiting for scientific consensus could in such cases mean allowing preventable disasters.) This Principle clearly coheres with the consequentialist approach to ethics presented in Chapter 4. One version of this Principle was included in the Rio Declaration of 1992, although that version merely made the absence of scientific consensus no reason to avoid action but implicitly allowed other reasons such as costs. Policies that could significantly undermine the habitats of wildlife would need to be rejected if this Principle is honoured. Several actual disasters (such as prescribing thalidomide to pregnant women) could have been avoided if this Principle had been internationally recognized earlier than it was.

Manifestly there are risks that some (perhaps many) of the SDGs will not be delivered by 2030, despite the progress made towards some of them at Paris (2015) and Kigali (2016). Trillions of dollars would be needed every year to attain these goals, and yet not many countries attain even the UN goal of 0.7 per cent of gross domestic product (GDP) allocated to foreign aid.

Further, some of the goals could be held to be insufficiently ambitious. An income of US\$1.25 per day has been retained as the threshold of absolute poverty, a threshold that many hold to be far too low; thus it is feared that the attainment of the first goal would not solve the problem of poverty, which is widely held to be a precondition of solving global problems in general. Yet international agreement about goals and targets is likely to galvanize far more effort and commitment than would arise in its absence. The adoption of these particular goals is foreseeably beneficial, and is thus (despite the various problems) to be welcomed.

Meanwhile the participation of individuals is being fostered through the ongoing 'Global Goals' Campaign. Participation in this is one of an increasing number and range of emerging opportunities for active global citizenship. Those who see themselves as global citizens recognize the rights of people everywhere, and that their own responsibilities straddle national boundaries. Examples of these responsibilities include joining in efforts to tackle environmental problems, many of which bestride frontiers (and species boundaries too). Even if other problems did not call for global citizenship, the worldwide and systemic nature of environmental problems makes the case for global citizenship inescapable.

Ecological preservation

Some readers who recognize the need for SDGs to end poverty and hunger may be more hesitant about goals to preserve biodiversity, even if they accept such environmental goals as the goals to limit carbon emissions and to replace energy generation from fossil fuels with electricity from renewable sources. So it is worth reflecting on what makes biodiversity loss a global problem, and what forms of preservation should be pursued.

The grounds of much preservation rests on the symbolic value of historical landmarks or artefacts (think of the Bayeux Tapestry), and similar grounds apply to the preservation of significant fossils such as archaeopteryx. But the grounds for preserving living species, sub-species, and the habitats on which they depend do not turn in the same way on human aesthetic responses or on historical or scientific interest, even though wonder at the natural world is a key motive, and perpetuating opportunities for future generations to share such feelings and responses itself constitutes one of the grounds for ecological preservation.

The extent and scale of biodiversity loss should first be remarked. Losses to biological diversity (animals, plants, and other creatures) have become so vast that the rate of loss may already be exceeding the rate of diversification implicit in the evolutionary process itself. Of an estimated total of nine million species, something like a quarter are at risk of extinction over the coming three decades. Since under two million have been identified, many could well be lost before even being noticed or recognized. Losses are particularly striking in vulnerable areas such as wetlands, estuaries, coral reefs, and rainforests, where species diversity is at its greatest. At the same time, deforestation is probably affecting global climate, and thus multiplying global climate change for creatures of every species.

Admittedly there are problems about the definition of 'species'; but working scientists assume that these problems can be solved. Thus on one definition, species are populations whose members are capable of inter-breeding and producing fertile offspring. While this definition does not work for species with asexual forms of reproduction, and turns out to supply sufficient but not necessary conditions for being a species, it is successful enough to show that species are not mere subjective constructs, but distinctive objective units of the evolutionary process of speciation.

Yet extinction rates of species are accelerating, and a million species may have been lost already. Relatedly, the extinction of any species involves the loss of the value which would have been carried by subsequent generations of that species, the lives of which are now pre-empted. In combination, all this shows species-loss to be a global problem, even without the human interest in species preservation being considered.

Nevertheless, the reasons for preventing species-loss should be further considered. Some thinkers take the view that the reason for preserving biodiversity is its aesthetic value for human beings. This is sometimes said to consist in the emblematic value of species such as eagles, which are held to symbolize American values, while others include the appreciation of clusters of species as experienced by eco-tourists. But these grounds at best justify localized preservation, and are prone to fluctuate with the waxing and waning of human tastes.

More impressive is the argument that compares living nature to a genetic library, and the destruction of forests to burning a library of volumes that remain unread. This is in part an argument from the value of scientific study, and the way that it adds to human understanding and flourishing.

It is also an argument from the uses that widely result from the study of wild species. Thus crop failures are sometimes overcome through the discovery of genes resistant to predators and carried by the wild relatives of food plants, such as the variety of wild maize (mentioned in the Brundtland Report) found in a Mexican forest under threat of destruction, a variety which could prove vital to the world production of maize.

Also a high proportion of pharmaceutical products have been discovered in the plants or other creatures of rainforests, and there is every reason to preserve such ecosystems, endangered species included, to allow the search for further remedies to continue. This is implicitly an inductive argument from the frequency of discoveries of remedies to the likelihood of further discoveries, if their possibility is not foreclosed. It also epitomizes arguments for preservation based on the 'ecosystem services' of natural systems to humanity.

A further argument relates to the dependence of humanity on nonhuman nature. Wild populations and species have been compared by Anne and Paul Ehrlich (1994: 335) to the rivets which hold together an aeroplane. Many rivets can be removed before the plane becomes unsafe, but it is unwise to rely on a plane from which rivets are regularly removed. Analogies are not arguments, but the multiple dependencies of humanity on nature suggest that this particular analogy upholds wise policies of preservation. To cite one of many examples, James Lovelock has discovered the production by the bacteria of estuaries and continental shelves of dimethyl sulphide (which regulates the proportion of sulphur in oceans) and of methyl iodide (which regulates the proportion of iodine). Myriads of living creatures, humanity included, depend on the ongoing generation of these regulatory substances.

This indicates the dangers of human interventions with the biota of continental shelves and estuaries (dangers which can be recognized whether we endorse Lovelock's planetary theories or not). Forests too, it turns out, are vital for regulating rainfall, absorbing carbon, and preserving levels of atmospheric oxygen (roles which have come to be known as 'ecosystem services'). Most living creatures turn out to depend on the intactness of such planetary systems. So the current argument is based not only on human interests but on those of the generality of living creatures.

These various arguments supplement arguments from the intrinsic value of the well-being of the creatures themselves, whether present or future, and help explain why biodiversity loss is a major global problem, and why its preservation warrants inclusion in the SDGs. This does not mean that there are no problems about the nature and extent of preservation, problems discussed in the following section, but it does indicate that the preservation of natural systems and of major habitats such as rainforests and coral reefs is vital, and needs to figure in any programme of sustainable development, whether grounded on human interests alone or (as has been advocated in earlier chapters) more broadly.

Forms and limits of preservation

The Convention on Biodiversity, initiated (as we have seen) at the Rio Conference of 1992, was furthered by an agreement made at Nagoya, Japan, in 2010, concerning genetic diversity, the regulation of genetic resources, and the restoration of degraded ecosystems (such as the Colorado delta in Mexico). It was also agreed then to enlarge funding by developed countries for biodiversity protection in biodiversity 'hot-spots' in developing countries. The subsequent SDGs concerned with 'life below water' and 'life on land' were endorsements (on the part of a larger number of signatories) of the Nagoya provisions.

But as John Passmore once remarked, not everything can be preserved, and so preservation has to be selective. While it is sometimes feasible to restore an ecosystem to its condition prior to human intervention, this aim is often unachievable, partly because human intervention has generated new ecosystems, with species dependent on (for example) grazing by domestic animals such as sheep, and partly because ecosystems are never static but constantly in a dynamic state of flux. (Thus restoring the forests of Hawaii to their pre-European or their pre-Polynesian condition turn out not to be practicable propositions.) While particular species can be restored to their historical habitats (such as sea eagles in Scotland and red kites in Wales), it is not feasible, even if it were desirable, to erase the impacts of human settlements and return those places to their condition at the end of the last Ice Age.

Critics of ecological restoration also claim that if the objective is the promotion of biodiversity, humanly modified areas are often at least as biodiverse as the corresponding pre-human ecosystems. While this is sometimes true, it is not always a conclusive reason to take no action. The biodiversity of some parks, gardens, and arboretums often warrants preserving their present condition (where possible) rather than their conjectural past. But where current biodiversity has resulted from the introduction of species (such as minks in England, rhododendrons in Scotland, eucalyptus in Africa, and wattle in Australia) which threaten the continued existence of the species present before their introduction, long-term biodiversity is best secured through attempting to remove them.

Not all alien species should be eradicated; for example, it would be pointless to eradicate culturally well-established species introduced into Britain such as rabbits and horse chestnuts. But sometimes alien invasive species (such as eucalyptus in Africa and snakehead fish in the USA, prone to undermining the creatures of native ecosystems) must be removed if future generations are to experience traditional ecosystems, and the range of biodiversity that they encompassed is to survive and have future generations of its own. The interests of these systemically related future generations outweigh in such cases those of future generations of invasive and disruptive species.

Restoration of forests can also be justified where human exploitation or conflict has led to their removal. Northern and central Ethiopia, for example, have lost most of their tree cover through civil war; in Haiti, however, deforestation has been due to exploitative farming. The efforts of these countries to re-afforest the affected areas should be applauded, whether the resulting woodlands closely resemble the woodlands of the past or not.

It should not be forgotten that one of the main causes of deforestation is conflict (often in the form of civil wars), and that another is exploitation by farming or mining (such as the illegal gold extraction that continues to affect much of Colombia). Warfare is among the biggest threats to natural systems. Resolving conflicts and curtailing exploitative forest clearances (as in the restoration of the cloud forests of Costa Rica) must figure prominently in attempts to preserve the world's remaining forests. 'Rewilding' them so as to reverse the impact of humanity will seldom be feasible, but allowing them to recover from the grosser scars on the landscape is more often attainable and worth attaining.

Other critics of ecological restoration urge us to accept what they call 'novel ecosystems': areas affected by human interventions but currently sustaining themselves without further human interference. This concept, however, is difficult to evaluate, as almost all the areas affected by human interventions (ice-caps excepted) continue to be managed or otherwise affected by humanity.

What we have to recognize is a whole range of humanly influenced ecosystems, from most of the Amazon rainforest (long since affected by its human inhabitants, but well worth preserving), to industrialized cities, brown-field sites, and derelict canals. While these cannot be restored to a pre-human condition, many can be enhanced so that cities include green spaces and urban agriculture, and enhanced waterways can be restocked with aquatic wildlife.

Sites such as these do not comply with the concept of novel ecosystems, but can serve the preservationist aim of re-introducing ecosystems that thrive through a combination of human management, natural processes, and human forbearance. Accordingly the earlier contention that ecological preservation must play a central role in policies of sustainable development turns out to be amply vindicated.

Chapter 6
Social and political movements

Deep Ecology

Large contributions have been made to environmental ethics by social and political movements. Some of these are briefly introduced here. While Deep Ecology has already been mentioned in Chapter 1, it is worth returning to it, to compare it with other movements such as ecofeminism, which has often been opposed to it, despite possibilities for joint campaigning. Social Ecology, the Environmental Justice Movement, and Green political movements will also be considered.

Deep Ecology commendably stresses the long-term, global, and inter-species aspects of environmental concern. The Norwegian philosopher Arne Naess was its most famous proponent, and has put forward what he calls 'the Deep Ecology Platform'. As we saw in Chapter 1, this platform favours equality within and between species, upholds diversity both of life-forms and of cultures, rejects all forms of exploitation, and supports the broadest possible interpretation of the fight against pollution and resource depletion. Further, it fosters human societies in which multiple forms of work are respected and integrated. This movement has found followers particularly in Australia and United States. It has also proved attractive to some adherents of James Lovelock's Gaia theory, for whom the Earth is a self-regulating and interconnected system; but

Deep Ecology advocates defending planetary nature everywhere, and not only the rainforests, estuaries, and continental shelves which Lovelock regards as distinctively vulnerable, and urges us to leave intact.

The value-theory of Deep Ecology is based on self-realization, which involves, according to Naess, our identification with other living beings. Our identity already, he claims, includes whatever we are related to, whether human beings or other species; and realizing our true selves involves expanding and merging what we see as our individual interests with those of other beings, however diverse, and reacting so as to defend them accordingly. Where ethics is concerned, it suggests that once this kind of identification is achieved, no further ethic or ethical reflection is supposedly needed.

Yet many people have an ethical concern to defend other people, other species, and ecosystems such as rivers and mountains, without identifying with them. This is all possible without seeing others as a greater self of which we are just an aspect. We are able to respect other people and significant places just because we have distinct identities. It is our own very distinctness that gives our ethical convictions their role. The kind of motivation stressed by Naess can be important, offering those who follow him crucial imaginative possibilities. But other kinds of motivation, not based on self-defence, but on respect for and love of other creatures, landscapes, and environments, can be equally important, and uphold the kind of ethics which can weigh up diverse and sometimes conflicting concerns and priorities, such as present interests and future ones, or those of different species.

Other problems arise about the aspiration of Deep Ecology to reduce the human population; but this topic has already been discussed in Chapter 1. What can here be said for the first time is that the kind of ethical reflection which was discussed in Chapter 4, and which Deep Ecology, through its claims about the sufficiency of identification, implicitly discourages, is all-important, both for shaping the human future and for the health of the planet. The hard work of ethical reflection should not be side-lined, attractive as the case for expanded identification may often be. For this reason, Deep Ecology should not be considered the last word in environmental ethics, even though it has opened the eyes of many to environmental problems and to social possibilities, both local and global.

Ecofeminism

Françoise D'Eaubonne devised the word 'ecofeminism' as long ago as 1974, for reflection and activism related to the intersection of feminism and environmental thought. In its early days, ecofeminism developed insights such as that of Simone de Beauvoir, who had earlier maintained that patriarchal (or male-dominated) systems treat women and nature alike as 'other'. These insights were taken further by Karen Warren, who stressed the links connecting exploitative relations between men and women and exploitative relations between humanity and nature. These, she claims, are closely associated forms of oppression, and neither can be overcome without due attention being paid to the other. While ecofeminism has presented important additional themes, this was its original central emphasis.

Others have pointed out that there are many forms of oppression and domination, including racism, classism, the exploitation of workers, and the persecution of religious and sexual minorities, as well as sexism and the human domination of nature. Equity might here suggest that they should all be tackled whenever they arise, and simultaneously if necessary. Ecofeminists (and Deep Ecologists too) would not disagree, but ecofeminists assert close historical connections between the oppression of nature and that of women in particular (and often berate Deep Ecologists for decrying the former while remaining relatively silent about the latter).

Connections of this kind have been affirmed by Carolyn Merchant, who contrasted pre-modern respectful attitudes to 'mother Earth' with early modern and subsequent advocacy of exploring nature's secrets (through mining and experimentation) and, equally, in the name of scientific inquiry, the practice of vivisection. Whether or not these historical changes of attitude bear out a conceptual connection between attitudes to women and to nature, or instead reflect persistent metaphors (like 'putting nature to the test') used to justify diverse and often exploitative practices, these historical links have been held to require that exploitation of nature and of women be considered and treated together.

But these two forms of oppression seem not to go together in every society. The philosopher Workineh Kelbessa relates that within his own society (the Oromo of Ethiopia), while women are often oppressed, nature and wildlife are not. This suggests that the linkage between these forms of oppression is confined to particular societies and epochs, at most. Yet it is arguably far from universal even in Western societies, where women often play a prominent part in oppressive practices like fox-hunting and other blood-sports, and to this extent figure among nature's oppressors rather than among fellow victims of oppression.

Nor are Western attitudes to nature uniformly oppressive. While many (of both genders) consume the products of factory-farms, many others campaign against this practice, and still more make strenuous efforts to protect wildlife. Thus claims about oppression should not be overgeneralized. While all kinds of systemic exploitation (including that of women) should be contested, there does not seem to be the strong systemic correlation between the exploitation of nature and of women that some ecofeminists claim.

Nevertheless, ecofeminists, in diagnosing these kinds of exploitation, have come up with valuable correctives to much previous thinking, not least about the environment. They have, for example, criticized an excessive emphasis on dualisms, and the kind of thinking that regards pairs of apparent opposites as mutually exclusive and conflicting. Thus male and female have often been treated as polar opposites, and nature and culture too, as if these categories had nothing in common. Much the same polarity has been assumed to apply to reason and emotion, reason being associated with masculinity and being valorized accordingly, and emotion with femininity and being correspondingly devalued. Ecofeminists including Warren have challenged such dualistic thinking, and have also suggested different approaches, intended to improve on them, not least in the field of ethics.

There is much to be said for ecofeminist objections to polarized thinking. Feminists rightly protest when kinds of work are stereotypically represented as men's work or women's work. Further, attitudes treating nature and culture as antithetical can produce such travesties as (on the one hand) urban contempt for rural life and (on the other) regarding only untouched wilderness and its creatures as valuable. (Some have even decried ecological restoration and its outcomes as defective and deceptive simply because they are dependent on human effort.) Yet the human engagement with nature largely takes place through farming and gardening, both of them aspects of culture, and refusal to recognize the dependence of both on nature and on natural processes frustrates both these activities, as well as diminishing our own sensibilities.

Ecofeminists have also valuably foregrounded the role of emotions such as compassion, and decried excessive emphasis on reason, not least in ethics. Val Plumwood has rightly stressed the importance of emotional sensitivity, particularly in relations with animals, and how reliance on reason and on principles alone (as in Kantian ethics) can fail to motivate the discharge of responsibilities that we intellectually endorse, and produce unnecessary self-division. At the same time, she has criticized instrumentalist and egoistic attitudes to everything other than the self, as detrimental to the kind of sensitive relations with the natural world necessary for its protection.

In a similar vein, Mary Midgley has criticized the kind of atomistic individualism which ignores both our complete dependence on others in infancy and childhood, and our willingness as adults to care for others, and which makes society a contract between rational but emotionally stunted individuals imprisoned in their own self-interest. (This criticism was echoed in Seyla Benbabib's critique of Rawlsian contractarianism: see Chapter 4.)

Yet others, in the light of such critiques as these, have advocated an ethic of care, rightly emphasizing that we learn to care within relationships. Such an ethic works best for roles within communities where responsibilities are reciprocal. But (as we have seen when considering responsibilities relating to people of the distant future) many of our responsibilities are non-reciprocal, while remaining valid and significant, and extend far beyond relationships actual or possible. Besides, the areas of morality concerned with caring are liable not to extend to further areas such as those of fairness, justice, and equity. So there are limits to an ethic of care, important as it often is, and an ampler ethic is required if, for example, the people of the next century are to be given due attention and significance.

However, the ecofeminist critique of polar thinking and of atomistic individualism comprises a major contribution to philosophy. It liberates us from individualist and contractarian understandings of society, and facilitates recognition of attitudes and emotions that make many of the virtues possible, and that patriarchal thinking all too readily ignores or suppresses. And without endorsing Deep Ecology, it helps explain the willingness to identify with other creatures which Deep Ecologists seek to instil. To its credit, it recognizes our embodied and socially embedded situation. Yet we should hesitate to regard all women as victims of oppression.

Nor, at the same time, should we underestimate the ability of women to influence and change the future of the planet.

Social Ecology

The Social Ecology movement was pioneered by the socialist Murray Bookchin, who regarded ecological problems, like other problems, as fundamentally social in nature. A strong case can in fact be made for the oppressive treatment of nature as an extension of the hierarchies of domination that have long blighted humanity, such as the oppression of one class by another, or discrimination on the basis of colour or gender. Bookchin's remedy consisted in the fostering of democratic decision-making and participation at all levels, the kind of proposal which many will find congenial as part of a solution to social and economic problems, together with some environmental problems such as emissions from vehicles and pollution of the air, the rivers, and the oceans.

There is room to doubt, however, whether this basically humanist approach has the potential to overcome the exploitation of animals, in view of their inability to raise voices of protest against their treatment in factory-farms and experimental laboratories. If it had, there would have been little or no need for the 'animal liberation' movement led by Peter Singer, or for campaigns like Tom Regan's advocacy of animal rights. People might have attempted to preserve 'ecosystem services' or to reduce their consumption of meat for the sake of their own good; but without independent concern for non-human interests, efforts to protect the natural world, wild species, and their habitats could well prove to be insufficient.

The dangers emerge more clearly in the light of Bookchin's suggestion that humanity should take charge of the progress of evolution through systematic genetic engineering. (Some advocates of the Anthropocene Age, mentioned in Chapter 1, echo this suggestion.) While there may be a place for selective genetic engineering (consistent with the precautionary principle), for example to avert malnutrition or famine, the idea that humanity might understand enough to take control of the evolutionary process in general would require a much greater grasp of biology and the good of the species to be engineered than is likely to be achieved in the foreseeable future. This suggestion amounts to advocacy of a domination over nature that is both dangerous and arrogant, just as it has been since Enlightenment thinkers first proposed it.

The Environmental Justice Movement

This is the name of a movement that campaigns against discrimination against disadvantaged groups or communities, for example with regard to exposure to radioactivity and the siting of toxic and other waste 'facilities'. Examples have included the contamination of Navaho lands in Arizona through uranium mining, and the exposure of Navaho miners there to levels of radioactivity far exceeding allowable limits. Elsewhere the planned Yucca Mountain high-level nuclear waste disposal site in New Mexico posed a threat to Shoshone and Paiute sacred lands, until this project was cancelled in 2012.

Earlier, in North Carolina, the citizens of Warren County, most of whom were African Americans, protested unsuccessfully against the siting of a polychlorinated biphenyl (PCB) dump in their community. This protest inaugurated the Environmental Justice Movement, and led to a study by the United Church of Christ Commission for Racial Justice (1987) disclosing that hazardous waste sites tend to be placed in areas with large minority populations.

This practice turns out not to be confined to the United States, but to have an international dimension. Western companies have often dumped toxic wastes at sites in West Africa, and electronic waste, full of heavy metals and other hazardous substances, has been exported to India, Africa, Bangladesh, and China. Lax regulation at such sites often allows children to use dangerous practices in attempting to recover saleable materials, at great personal risk (as Marion Hourdequin relates). When the dumping of dioxin-laden industrial ash from Philadelphia in Guinea and Haiti (1987) and of PCB-contaminated chemical waste from Italy in Nigeria (1988) are added to the record, this practice turns out fully to deserve the label attached to it by James Sterba: 'environmental racism'.

The Environmental Justice Movement objects to unfair distributions of environmental harms such as pollution, and to inadequate procedures, as where people lack a say in decisions affecting themselves and their locality. (These are, respectively, issues of substantive justice and of procedural justice.) Thus Sterba has proposed a 'Principle of Procedural Justice', by which 'Everyone, especially minorities, should participate in the selection of environmental policies that affect them.'

But beyond procedural justice, there are also issues of recognition, as when minority peoples, because of patterns of cultural domination and disrespect, are effectively ignored, despite provisions of procedural justice intended to include them. Recognition is a worldwide problem, alongside substantive and procedural justice, and needs to be taken as seriously. Besides, there can be a lack of recognition at the level of communities, as when food insecurity afflicts communities of farmers through the unintended impacts of policies of multinational companies.

Environmental injustice in Africa is not confined to the dumping of toxic wastes on Africa's western coast, as Kelbessa points out. Containers full of toxic waste were dumped a short distance off-shore along over 400 miles of the eastern coastline of Somalia, when it had no government to object. Then waves from the Sumatran tsunami of 2004 broke open many of these containers and scattered their contents (including radioactive materials and heavy metals) over the surrounding area, causing untimely deaths and probably some of the local cancer clusters. These harms are set to continue until preventive measures can be initiated by the recently installed government. This terrible story exemplifies issues of substance, procedure, and recognition alike.

The Environmental Justice Movement, then, turns out to have worldwide implications (as the Bhopal disaster serves to confirm, and likewise the global spread of oil-spills), and raises important issues of justice between members of the current human generation, which bring out the need for principles of compensation as well as of distribution, procedure, and recognition. Whether the principles and practices needed to resolve these issues (compensation included) can be fitted into the kind of consequentialist framework commended in Chapter 4 is for the reader to judge, and must be set aside here. It must be clear, though, that the overall impacts of such remedial practices must be taken into account in appraising them.

The more immediate question concerns the relation of this movement to environmental ethics and its scope. One answer is that the insights of this movement must not be forgotten when issues of contemporary justice are being considered, even though there are contemporary inter-human issues such as inter-state relations on which it has little to say. Another is that justice between generations matters as well as justice within generations (see Chapter 3). This was recognized in the principles adopted in 1991 by the First National People of Color Environmental Leadership Summit, and is routinely stressed by indigenous campaigners in many countries. Other parts of this movement, however, may sometimes be in danger of failing to emphasize issues of inter-generational justice.

Yet another is that this movement shares some of the limitations of the Social Ecology Movement in being confined to human interests, and that unless non-human interests are taken into account, decisions could well fail to be right or just. For justice cannot reasonably be restricted to inter-human dealings alone. Nevertheless, the movement has drawn attention to forms of oppression and discrimination which had previously been largely overlooked, and which any satisfactory environmental ethic needs to highlight and seek to halt.

The Green movement

Green political movements have prioritized various themes of the movements discussed in this chapter, together with policies of sustainability, climate change mitigation, and adaptation, and of resistance to pollution and polluting processes. In the light of humanity's carbon budget (see Chapter 8), they characteristically support energy generation from renewable sources, and oppose the mining and extraction of fuel, particularly through new technological processes such as fracking, holding that coal, gas, and oil are best kept in the ground. Some of their members adhere to stances that Naess would consider 'Deep', others stances he called 'Shallow', and yet others a whole variety of intermediate positions. Importantly they are widely prone to oppose the assumption that economic growth is to be welcomed.

This is not the place to survey the fluctuating fortunes of Green parties, whether in Britain, other European countries, or elsewhere in the world, nor their detailed economic policies, nor the alliances that some have formed to take part in governmental coalitions. But it is worth considering how self-proclaimed environmentalist parties have been able to participate in the democratic processes of (more or less) liberal and market-oriented societies, with the associated need to appeal to their electorates. To what extent can environmentalism be reconciled with 'liberalism' and with growth-oriented economies?

Konrad Ott distinguishes four forms of opposition to growth. The first rejects treating GDP as the criterion of success and national well-being, preferring the goals of quality of life and of happiness. Many organizations, including most Green parties, would endorse this approach, which does not insist on negative growth-rates, but accords growth-rates secondary or subordinate importance. (Meanwhile countries such as Bhutan and Costa Rica claim far to outstrip more developed economies in their happiness levels.)

A second variety seeks to reduce the impacts of growth on natural systems, emphasizing strong sustainability, and simultaneously fostering sustainable development in developing countries. As mentioned in Chapter 5, the pursuit of strong sustainability diverges from maximizing benefits for humans, sometimes on non-anthropocentric grounds. Efforts to preserve species and wild habitats comply with these aims, and enjoy wide and growing support. A possible future development of this variety would aim at greater global equality.

A third type seeks to reinstate communal conviviality, upheld by the kind of virtues commended by virtue ethicists. The suggestion is that in such societies the foregone gains of growth would not be missed. Yet, we might comment, they might well be missed by the poor and the disadvantaged, unless special provision is made for these groups at the same time.

A fourth type takes the view that capitalist modes of production and distribution are incompatible with all the varieties of de-growth mentioned here, and need to be replaced with cooperative structures. But a variant of this type takes the view that such changes would risk losses of liberal freedoms, and that particular moves away from capitalism therefore need to be carefully evaluated. An argument supportive of the suggested incompatibility is that the growth inherent in capitalism cannot continue forever, and must eventually come up against limits to finite resources. A reply is that the incompatibility is with *unbridled* capitalism, whereas forms of liberal society that recognize limits to growth, regulation of capitalist enterprises, and goals such as quality of life and strong sustainability remain possibilities, whether introduced by Green parties or by others.

These considerations bring to the fore the extent to which political environmentalism is compatible with liberal democracy. The tensions are between the kinds of regulation of production and consumption that consistent environmentalists are committed to favour to attain strong sustainability and preserve the natural world, and the liberal belief in individual choice.

Some forms of liberalism insist on market economies being untrammelled. But there are other forms, such as that of John Stuart Mill, which recognize limits to growth, and goals such as the preservation of wildlife and of related habitats (whether for the sake of our successors or of wildlife itself). These forms are less intransigent, while remaining committed to liberal freedoms such as freedom of speech. In exercising the liberal right to vote, people are free to support this kind of liberalism, and in this way the tensions are capable of being overcome.

One example of possible conflict between liberalism and environmentalism concerns automobile use and ownership. The congestion and pollution caused by the use of cars are becoming intolerable, particularly when we consider both emissions contributing to global warming and those (such as nitrous oxide and particulates) that undermine air quality and contribute to disease and premature death. Many democratic authorities have been led to consider either restrictions on car use in certain areas, or subsidies for the scrapping of older and polluting cars. While motoring organizations uphold drivers' freedoms, and press for more roads, it is by no means clear that such liberal freedoms should have the last word, or that Green advocacy of walking, cycling, electric cars, and use of public transport should not be heeded.

A comparison and overview

Social Ecology and the Environmental Justice Movement serve as correctives to Deep Ecology in foregrounding the social structures in which environmental problems are often found. But Deep Ecology and many ecofeminists serve as correctives to these movements in their turn, with their concern for non-human creatures, their species, habitats, and ecosystems.

Ecofeminists add the importance of avoiding polarized thinking and of contesting oppression in all its forms, and their understanding of human agents as embedded in relationships, and as responding to nature and society through emotional ties as well as reasoned principles. The Environmental Justice Movement stresses environmental injustices both within and between societies, and reminds us not only of people's right to be recognized and consulted about their environment, but also of the importance of present compensation for past inequities, and of future compensation for present ones. Green movements (and Deep Ecology too) contribute an emphasis on our obligations to future generations and to the non-human world.

The opposition of Greens to growth can assume different degrees of plausibility, with their stress on quality of life rather than GDP and on strong sustainability among their more defensible themes. Tensions can arise between environmentalism and liberalism, but they are not always insuperable.

Chapter 7
Environmental ethics and religion

Lynn White's critique

This chapter opens with a discussion of a widely influential attack on Judaeo-Christian attitudes to nature which finds in them the roots of our ecological problems, and proceeds to expound and defend the stance of stewardship adopted not only by Christianity but also by Judaism and by Islam. Given that the cultures of approaching half of humanity are strongly influenced by one or other of these three theistic faiths, and that secular versions of stewardship are available (arguably in a defensible form) to people from these cultures who have abandoned religious belief, and to people from other cultures too, a focus on theistic religions and the related tradition of stewardship is indispensable in an introduction to environmental ethics (however short). Nevertheless the other religions of the world have made their own contributions to the care of the environment, and the chapter concludes with a survey of some of the more striking of these contributions.

It is widely believed that Western forms of religion have fostered an anthropocentric attitude to nature, and with it a despotic and domineering approach. At the same time, millions of people in the West and elsewhere believe that humanity's relation to nature is that of a steward or trustee, many of them holding that in this role they have ethical responsibilities and are answerable to God for discharging them.

In an influential essay of 1967, Lynn White Jr argued that Christianity is essentially human-centred and committed to an arrogant and despotic attitude to the natural world, holding that it is God's will for humanity to exploit the Earth. White was an expert on medieval technology, and presented the introduction of heavy ploughing in the 7th century as epitomizing Western Christian arrogance. Humanity, he claimed, became at this stage the exploiter of nature. Yet he exempted Eastern Christianity from these charges. His remedy was the adoption of Zen Buddhism or, failing that, embracing the attitudes to nature of St Francis.

White's essay was much anthologized, and his message was widely disseminated in popular literature. This makes it worth saying that his main book on medieval technology took a much more nuanced and qualified approach to the same topics. Heavy ploughing, he relates in his book, began in the pagan ancient world, and was spread, among others, by the Vikings before their adoption of Christianity. White's essay, however, omitted such nuances and qualifications.

Other commentators hold that medieval Christianity was used to bless and justify technological advances in the West (like heavy ploughing) that were happening anyway, driven by economic rather than theological pressures. The exploitation of nature, begun in ancient times, was intensified in that period, but Western religion can hardly be held to be its origin or driving force.

Later, early modern Christianity certainly encouraged the scientific quest for laws of nature (think of Kepler, Galileo, Boyle, and Newton), as also had medieval Islam, to discover the creator's plan. But none of this makes Christianity human-centred, let alone commending the ruthless exploitation of nature for human purposes. The Old Testament has prohibitions against maltreatment of domestic animals (Proverbs 12: 10) and taking mother birds from their nests (Deuteronomy 22: 6–7), and recognizes that God has created times and places for wild creatures such as lions and sea-creatures (see Psalm 104 and Job 38–41).

Jesus urges his followers to consider the birds of the air and the lilies of the field (Matthew 6: 26–9), despite the greater value of human beings, and Paul has a place for the whole creation in God's plan of salvation (Romans 8: 21–2). In short, Christianity cannot fairly be represented as anthropocentric, let alone as favouring human exploitation of nature or despotism over other species.

As for White's selective commendation of St Francis, Susan Power Bratton relates that this saint 'far from standing alone, is only one figure among a fully developed tradition of Christian appreciation of nature as God's Creation'. As will shortly be seen, there is a Christian (but not solely Christian) tradition of a much more benign attitude to natural creatures. Nor is it likely that the adoption of Buddhism in the West would make a crucial difference; for, as John Passmore has remarked, Eastern religions have not in practice usually prevented the degradation of the environment in the countries of their greatest influence. And as Passmore further maintained, ethical traditions can only develop where what he called their 'seeds' (or incipient traditions) exist already, seeds of a kind which he discovered within Western traditions.

Nevertheless White's article stimulated theologians to develop what is often called 'ecotheology', or a theology of nature and of humanity's obligations towards the planet and its creatures. It also led to studies of the teachings of the various religions about nature and ecology. Indeed, White was later able to claim credit for making these developments possible.

Stewardship

Among the Western traditions depicted by Passmore is that
of stewardship, for which humanity is answerable for the
conservation and care of the natural environment. Passmore
considered this tradition as latent between pagan pre-Christian
writings and its assertion or re-assertion by Sir Matthew Hale
(17th century). But his mentor Clarence J. Glacken regarded it as
the general stance of the Old Testament, from the command to
'dress and keep' the Garden of Eden (Genesis 2: 15) onwards. He
also shows commitment among church fathers such as Basil,
Ambrose, and Theodoret to another benign tradition mentioned
by Passmore, for which the role of humanity is to enhance the
beauty and fruitfulness of the world of creation, thus completing
the creator's work. The monks of the order of St Benedict, for
example, often saw themselves in this light.

Glacken's interpretation of the Old Testament has in effect been
endorsed by Jewish writers like Jonathan Helfand, who stresses
that 'The Earth is the Lord's' (Psalm 24: 1). He accepts that for
monotheistic religions such as Judaism nature is not sacred,
but insists that, for Judaism, humanity is responsible and
answerable to God for its use. This raises the question of whether
the stewardship tradition is capable of being secularized, and
appropriated by people and institutions that have discarded belief
in God. That is a question to which we shall return.

For Islam, as S. Nomanul Haq relates, the world was created for
humanity, but for all generations and not just one. Humanity is
God's *Khalifa* (deputy), and human beings are thus global
trustees, accountable for their treatment of nature and other
creatures; hence mistreatment brings punishment. While the
Qur'an makes nature subject to humanity, it does not grant
unbridled exploitative powers over it, for it ultimately belongs
to God. Within Islamic tradition (*Hadith*), there is provision

for recognizing *hima* (protected pasturage, with special protection for resident flora and fauna), and also for *harun*, sanctuaries where killing animals of game species is forbidden, and where springs and water-courses are respected.

Fazlun Khalid has related (in an address given at Cardiff University in 2013) that as soon as Qur'anic insights about responsibility for the environment were translated into Swahili in 2001, and shown to the fishermen of Zanzibar, they immediately abandoned their long-standing practice of dynamiting coral reefs. (See Figure 5.) This practice was known to be illegal. But to them disobedience to the state was one thing; disobedience to Allah was quite another.

Thus while some other religions have regarded nature as sacred, the three great monotheistic religions (Judaism, Christianity, and Islam) have authorized its study and its use to satisfy human needs, but have also conferred on humanity the role of stewards

5. Fishermen off the East African coast blast their catch out of the water; blast fishing illegally harms coral reefs, contrary to Islam.

or trustees of the natural world. This is a tradition of which Passmore commends a secular continuation. It remains to be seen whether stewardship is open to any of the many objections that have been raised, and whether secular stewardship is possible.

Criticisms of stewardship

While stewardship is widely accepted in one form or another, numerous criticisms have been mounted against it. Several of these are considered here.

Sometimes stewardship is considered to be indelibly associated with its ancient and medieval origins, where the role of stewards included the supervision of slaves or serfs. To this criticism, Jennifer Welchman has replied that we do not dismiss democracy because it too originated in a slave society (ancient Athens). Another example of a practice that has outlived its social origins and is not regarded as besmirched by them is philosophy itself. The absence of underlying links between stewardship and evils such as oppression and sexism is borne out by the widespread alternative names used for this practice, 'curatorship' and 'trusteeship'. (Significantly, the goods of which curators and trustees have charge have a value recognized to be more than instrumental.) The response given here is relevant to stewardship in its secular as well as its religious version.

Stewardship is also held, because of its religious origins, to prevent respect for the natural world. The very distinctness of God as creator from nature is taken to detract from its value, by comparison with pantheism, for which God is co-extensive with the world. White's aspersions on Judaeo-Christian theology have sometimes stimulated this objection.

Belief in creation certainly advocates worship of the creator and not of creatures, but it also involves regarding the world as an expression of God's creative purposes, and God as indwelling the world (rather than absent from it, as some critics suggest). It requires human beings to respect nature as God's creature(s), and other creatures as fellow-creatures. For pantheism, by contrast, God is material, and there is no creator to whom worship and service are owed, and no creative purposes either.

While the Old Testament confers on humanity 'dominion' over nature (Genesis 1; Psalm 8), dominion is misinterpreted if it is taken to authorize human domination. Rather the commandment to take care of the garden in which the first people were placed can be understood as requiring a responsible and answerable attitude, one of stewardship. And while these are primarily replies that defend stewardship in its religious form, they also suggest that its secular versions, for which the language of creation and creatures is at best a metaphor, are immune from parallel objections despite their frequent indebtedness to religious language.

Many critics claim that stewardship is invariably anthropocentric. Some varieties certainly have been so, including those of Jean Calvin and of Islam (which, however, also require respect for God's creatures). But adherents of stewardship have often been non-anthropocentric; examples include such church fathers as Basil and Chrysostom, and such early modern figures as Sir Matthew Hale, John Ray, and Alexander Pope.

Non-anthropocentric stewardship has been advanced in recent times by the General Synod Board for Social Responsibility of the Church of England, and by such secular writers as Richard Worrell and Michael C. Appleby, whose definition is worth quoting in full:

> Stewardship is the responsible use (including conservation) of natural resources in a way that takes full and balanced account of the

interest of society, future generations, and other species as well as private needs, and accepts significant answerability to society.

This definition, besides disarming those who represent stewardship as invariably anthropocentric, also suggests an answer to the frequently posed question to what or to whom secular stewards are answerable.

Other critics accuse stewardship of managerialism, suggesting that it involves human interference with the entire surface of the planet to enhance the productivity of nature's resources. So stewardship is charged with an instrumentalist attitude to nature, and of adopting a managerial model. As Clare Palmer concludes,

> Stewardship is inappropriate for some of the planet some of the time, some of it for all of the time (the deep oceans), and all of it for some of the time—that is, before humanity evolved and after its extinction.

But there is no need for adherents of stewardship to adopt an instrumentalist attitude to nature, particularly when many Biblical passages appear to recognize its intrinsic value, and many Qur'anic passages resist such an attitude. Further, recognition of non-instrumental value involves respect for other species and their habitats, and thus refraining from colonizing the entire surface of the planet. Besides, stewardship is far from synonymous with interventionism, and is consistent with letting-be, appropriate for Palmer's own example of Antarctica, among many other places.

While Palmer rightly holds that there was no human responsibility before humans evolved, and that there will be none after human extinction, responsibility remains possible for the entire sphere of nature that human beings can affect, including nowadays the deep oceans, the solar system, and much of outer space beyond it. Unless the increasing human power is exercised with responsibility, global problems will be intensified. Thus, far from stewardship being inappropriate for any regions open to human impacts, human technology makes stewardship indispensable. It was, indeed, the very arrival of humanity on the planetary scene that made stewardship both possible and necessary. These defences apply to religious and secular stewardship alike.

James Lovelock has suggested that stewards will be prone to reach out for technological solutions to environmental problems, such as geo-engineering to solve the problem of climate change, and will support, for example, saturating the oceans with iron chloride to fix surplus carbon dioxide through the growth of algae. But this 'gunboat diplomacy' approach (as he calls it), or techno-fix mentality, conflicts with recognizing the Precautionary Principle, which environmental stewards are likely to favour without inconsistency. For his own part, Lovelock advocates seeing ourselves as planetary physicians (rather than stewards), taking steps to protect vulnerable species and mitigate greenhouse gas emissions. The role of planetary physician, however, is consistent and co-tenable with the role of steward, as long as neither is construed as involving planet-wide interventionism.

Yet further critics of stewardship maintain that it is liable to ignore social and international justice, and concentrate instead on the management of time, talent, and treasure. Some adherents of stewardship may succumb to this temptation, but they would, if so, be forgetting the ethical basis of stewardship, which includes, at least in the definition of Worrell and Appleby, fairness between generations and between species.

The appeal to ethics of these authors' secular stance is more than matched by that of Pope Francis, who in his encyclical *Laudato Si'* (subtitled *On Care for our Common Home*), advocates biodiversity preservation and urgent action on climate change, regards access to water as a human right, seeks to reduce inequality, and urges everyone to approach nature and the environment with wonder, in the spirit of his namesake, St Francis. Earlier Bartholomew, the Ecumenical Patriarch, had advanced a similar interpretation of Christian ethics.

Stewardship comprises a broad ethical platform, and is neutral between the various forms of normative ethics, such as virtue ethics, deontology, and consequentialism. So it cannot be expected to generate precise policy directives. But it is clearly committed to care for the environment and opposed to exploitation and environmental degradation, and thus guides attitudes and cannot be accused of being innocuous or vacuous. It also presupposes, in recognizing that everyone has a stewardly role, that those prevented by poverty or a hand-to-mouth existence should be put in a position to play their part. Thus the commitment of those adopting the stewardship approach to justice must mean enhancing the agency of the poor, so that they can join with the rest of humanity in the shared role of stewardship.

This conclusion presupposes the coherence of secular forms of stewardship, since religious forms are unavailable to many. But the main objection to secular stewardship, that it lacks answerability, has already been answered: secular stewards are answerable to society (both locally and globally). It is sometimes suggested that they are also answerable to future generations, but answerability to those who do not yet exist makes no sense. However, we can be answerable to our living children and grand-children, and also to the entire community of moral agents, the community which shares responsibility for caring for and preserving the environment, or rather to its present members, who are the ones in a position to hold us to account when environmental responsibilities are shirked.

Thus secular stewardship proves to be just as defensible as stewardship of the religious sort. It can thus be adopted by those who have discarded religious affiliations and by adherents of non-Western religions alike.

Contributions from other religions

But stewardship does not exhaust the contribution of religion to environmental care and protection. Other contributions include, for example, celebrations of nature, particularly when they foster love of the natural world and care for it. While many such celebrations take place in majority Christian, Jewish, or Islamic countries, others are inspired by different religious cultures and in lands influenced by them.

For example, in Japan the 'hanami' festival of flower appreciation spreads across the country as cherry trees burst into blossom, beginning in Okinawa and spreading gradually northwards across the country to the island of Hokkaido. So popular are these events that weather forecasts supply bulletins on when the blossom is expected to open. The short-lived blossom symbolizes the transience of life in Buddhist tradition, but the festivals also serve to foster love of natural beauty. As such, the festival is observed in the Japanese diaspora as far afield as the São Paulo province or Brazil, and in Vancouver and Toronto, Canada.

There are many corresponding festivities upheld by other religious cultures. No attempt is here made to be comprehensive, whether about celebrations or other traditional practices, as opposed to presenting a selection of environmentally promising customs or attitudes, often corresponding to the 'seeds' of benign Western traditions of Passmore's account.

In Chinese culture, Confucianism encourages its followers to become *ren* (virtuous or righteous), both in their actions and in their relationships. While, like the virtues of virtue ethics, this teaching is sometimes given an anthropocentric interpretation, it is capable of being extended to concern for non-human creatures (an extension that contemporary Confucians are free to consider). However, Daoism is more explicit in being environmentally benign. It rejects sharp distinctions between humanity and the rest of nature, and, according to Lao Tzu, teaches the equality of all creatures.

Marion Hourdequin draws attention to the early Daoist Zhuangzi, who imagines a fish becoming a giant bird, to the amazement of smaller creatures, and thus encourages us to change our conventional perspective and envisage new ones. In this way, the *dao* of Daoism diverges from that of Confucianism, all along encompassing the natural world and non-human perspectives, and implicitly fostering a broadening of the relationships of the virtuous person beyond relationships between human beings. This, Hourdequin suggests, opens up the possibility of an environmental ethic based on a relational self, but, unlike Deep Ecology, retaining respect for the different identities and perspectives, and the independent value, of other creatures.

Meanwhile, Simon James has commended as an environmentalist virtue compassion of the kind advocated by Buddhists. According to the Buddha, what is wrong with the world is *dukkha*, that is, suffering, understood as including dissatisfaction and cravings. *Dukkha* can be overcome through adopting an enlightened or better way of life, which will include a certain kind of compassion, one of the key Buddhist virtues.

This kind of compassion is compatible with being dispassionate with regard to *dukkha* affecting oneself, and requires overcoming self-centredness and becoming selfless. It also involves a heightened awareness of the suffering of others, without being overwhelmed by it. Such compassion (in this respect like ordinary compassion) is free from condescension, and can thus be contrasted with pity. But, often in contrast with ordinary compassion, it involves concern for the suffering of non-human animals (invertebrates specifically included); those lacking this kind of compassion are not, for Buddhists, compassionate at all.

Not everyone will agree that all dissatisfaction is to be shunned, nor that its presence should invariably prompt compassion. Discontent can sometimes lead to awareness of injustice, and to efforts to overcome it; and to this extent it can actually be welcomed. But there can be little doubt that compassion in the ordinary sense is a virtue that makes an overall positive difference to the world; and this verdict is not affected when the Buddhist requirement is added that animal suffering be included in its scope. If so, then non-Buddhists should consider broadening the scope of their compassion.

Perhaps more importantly, Buddhist teaching about *dukkha* and compassion enables Buddhists to develop their compassion so that it includes not only current suffering (whether human or non-human) but that of future beings and generations as well. In this way, the distinctive religious tradition of Buddhism can (and often does) facilitate benign attitudes to the natural world, its species, its habitats, and its climate.

Indeed in most if not all cultures, there are ethical and/or religious traditions capable of being developed to stimulate environmental consciousness. Thus among the Oromo of Ethiopia it is held to be wrong to destroy a species, as it would irretrievably reduce the creation of God (Wakka), even though reducing the numbers of harmful animals is acceptable. Besides, the Borana Oromo deliberately leave drinkable water close to wells for wild animals to drink in the night, believing that drinking water is among their rights. Practices of this kind admit of being broadened to a wider concern for wild nature.

Nor is the notion of sanctuary restricted to Western and Islamic religions. It is also to be found among the Venda, a people living in the Limpopo province of South Africa. Their sanctuaries are protected by guardians, who are themselves forbidden to harvest the fruits growing on the trees of these sacred sites. This traditional notion of protected areas has recently been spreading to biodiverse areas of land and of ocean, and has an important contribution to make to species and habitat preservation.

There again, the distinctive Bantu concept of *ubuntu*, or togetherness, advocated by many African sages, implying as it does that to be human is to be in a network of relationships, is widely held capable of fostering the kind of spirit of community needed to prevent environmental degradation and to alleviate the problem of climate change, particularly if it is extended to the biotic community, as has been proposed by Mogobe B. Ramose. This approach stresses both entitlements and obligations, extending potentially to the whole of humanity and to the whole biotic world. While not all sages adhere to such beliefs, many African leaders, including Archbishop Desmond Tutu, have suggested that this is an African contribution that can beneficially be shared with the people of other continents.

Elsewhere, some ecologically positive and widely publicized statements were presented in 1972 as the words of the 19th-century indigenous American Suquamish Chief Seattle. Speaking of his God and the God of white Americans, he appeared to have said 'Our God is the same God', going on to ask 'How can you buy or sell the sky?' and to declare 'The earth does not belong to man. Man belongs to the earth.'

Unfortunately, the director of the Southern Baptist film that presented these words omitted the acknowledgement on the part of their author, Ted Perry, that they were merely attributed to Chief Seattle. In the first written version of Chief Seattle's remarks, very different words were used, such as 'Your God loves your people and hates mine'. The film text probably made more impact through seeming to present nature mysticism from an indigenous sage; but care is needed when historical figures have contemporary (and possibly alien) thoughts placed in their mouths.

Nevertheless the words of Black Elk, from another indigenous American people, the Oglala Lakota (Sioux), were published in 1973, and appear to express the beliefs of many indigenous American cultures.

> We should understand well that all things are the works of the Great Spirit. We should know that He is within all things: the trees, the grasses, the rivers, the mountains, and all the four-legged animals and the winged peoples...and we should understand that he is above all these things and peoples....When we do understand all this deeply in our hearts...then we will be and act and live as he intends.

To add a contemporary example, the Haudenosaunee (or Iroquois) hold that 'There is a Creator who produced the things that give bounty to this life', that 'There is a living spirit in all things—animals, plants, minerals, water, and wind', and that people should 'live in harmony with nature' as well as with each other.

Thus the creator indwells creation, creatures deserve respect, and religious devotion can be the key to protection and conservation. These thoughts would be welcome to many adherents of other Western religious traditions, and cohere well with religious versions of stewardship.

Chapter 8
The ethics of climate change

Climate change and ethical principles

Urban pollution, oceanic pollution, species and habitat loss, and the growth of deserts are all serious ecological problems. But climate change in the form of global warming is almost certainly more serious still. Besides increasing concentrations of greenhouse gases such as carbon dioxide and methane in the atmosphere, it involves rising sea-levels, the flooding of islands and coastal settlements, the shrinking of glaciers and ice-caps, increasingly frequent and intense weather events such as hurricanes, droughts, and wildfires, the migration of millions of affected people and of numerous species to less inhospitable habitats, and the threat of worse such problems for their and our descendants.

There is little disagreement about the reality of increases in levels of carbon dioxide and other 'greenhouse gases' in the atmosphere. Carbon dioxide levels have risen from around 280 parts per million (ppm) in the pre-industrial period to around 400 ppm in the present. The level of greenhouse gases is somewhat higher, since (for example) methane and HFCs have many times the warming effect of carbon dioxide. 2016, 2015, and 2014 turn out to have been the hottest years in terms of average temperatures since records began. Average temperatures may not have risen quite as much as had been feared, but have still risen by 1 degree (Celsius) above pre-industrial levels, and are on course to rise much more if preventive action is not taken.

Besides, while not quite everyone agrees that human activity is the main cause of these increases, the vast majority of scientists agree that this is overwhelmingly likely, and that global warming is 'anthropogenic' (caused by humanity). This belief is reflected in reports of the Intergovernmental Panel on Climate Change (IPCC). In 1995, IPCC affirmed that human responsibility for global warming was 'more likely than not'. By 2001 they declared it 'likely', and by 2007 'very likely'. By 2013 they concluded that it is 'extremely likely'.

At the same time, their predictions of average temperature increases above pre-industrial levels range from 1.5 to 4.5 degrees (Celsius), with a small but significant possibility of increases being yet greater still. Yet increases of above 2 degrees could well be catastrophic, which is why it was agreed at the UN Paris summit of December 2015 to limit average increases to 2 degrees at most, and to 1.5 degrees if possible.

Hence, unless vigorous and concerted action is taken, there is a significant risk of human activity generating catastrophic climate change, catastrophic both for future generations, numerous species, and human victims of flooding and other extreme weather events in the present, including people who have made little or no contribution to causing climate change. There is thus a strong ethical case for vigorous and concerted action to mitigate climate change, and, given that some climate change is already irreversible, to adapt to its effects.

However, a small minority of scientists (and journalists and politicians too) persist in denying that climate change is anthropogenic. What should ethicists say in face of such scepticism? A good answer lies in the Precautionary Principle, which specifically applies wherever complete scientific consensus is lacking. As mentioned in Chapter 5, this widely endorsed principle advocates action to prevent outcomes from which there is reason to expect serious or irreversible harms, even in advance of scientific consensus. But sceptics cannot seriously deny that there is reason to expect outcomes that will be both serious and irreversible (if not prevented), despite denials from the sceptics that there is conclusive reason to expect all this.

So everyone (sceptics included) who accepts this principle should also accept the ethical case for vigorous and concerted action. This verdict is confirmed if you glance at *The Atlas of Climate Change*, incorporating chapters like 'Disrupted Ecosystems', 'Threats to Health', and 'Cities at Risk' (New York, Los Angeles, Mumbai, Shanghai, and Tokyo among them; but for the Thames barrier, London would be another).

Further ethical principles are also relevant. For example, it is reckless and wrong to inflict harms (such as the impacts of severe weather events) on people who have played little or no part in contributing to their genesis, as was pointed out by victims of hurricane Irma, when it recently struck the West Indies (September 2017).

There again, it is wrong avoidably to lower the quality of life of future generations, as our generation is doing through its failure to mitigate greenhouse emissions (as was argued in Chapter 3 on ethical issues relating to future generations). Future generations are likely to be subject to more intense and more frequent extreme climate events (droughts, hurricanes, floods, and wildfires), to the spread of previously tropical diseases like malaria, and to rises in sea-level liable to inundate coastal settlements and entire island territories, unless we act to prevent these impacts of our emissions. (In 2009, the government of the Maldives drew this matter to world attention by holding a cabinet meeting under water: see Figure 6.)

6. Underwater Cabinet Meeting, Maldives, 2009, symbolizing risk of wholesale inundation due to rising sea-levels.

Besides, it is wrong, as we have seen, avoidably to eradicate species except to prevent yet greater evils. Thus, the ethical case for strong and concerted action to mitigate climate change, to foster adaptation where it is irreversible, and, where possible, to compensate its victims is unanswerable.

Even if there were universal agreement about mitigation, adaptation, and compensation, the policies required are not immediately obvious. For there are other problems to consider at the same time, such as the needs of people and countries afflicted by poverty, problems of feeding a growing population, and ecological problems other than climate change.

Thus, issues of adaptation and of compensation have to be considered in conjunction with policies of sustainable development. Issues of food production and of the availability of fresh water need to be remembered, together with the desirability of empowering and educating women and thus stabilizing human numbers as soon as possible. Meanwhile the case already presented for preserving species and habitats needs to be heeded, together with avoidance of land degradation and the pollution of earth, air, and oceans. Integrated policies, both local and global, will be needed, challenging as this may prove.

Here, however, the emphasis will be on introducing (both locally and globally) policies of mitigation, adaptation, and compensation. For there are different defensible approaches to these issues, and it is time to consider them.

Entitlements, responsibilities, and regimens

How should entitlements to emit greenhouse gases and responsibilities to pay for mitigation, adaptation, and compensation be allocated? More than one suggested form of allocation on a global basis has been put forward.

Many (including Dale Jamieson, Peter Singer, and myself) have been attracted by a system called 'Contraction and Convergence'. The underlying principle is that each person has an equal right to emit greenhouse gases to everyone else (whether directly or from domestic animals used for transport or food). So the permissible total of emissions for a given year should be calculated and then shared among the various states of the world in proportion to their human population.

Countries wishing to emit above their entitlement would have to purchase some of the quota of countries not using their full entitlement; in this way the scheme would be redistributive, supplying poor countries with additional resources. Increasingly the criteria for entitlements would diverge from current practice and converge globally (Convergence), while the permissible total would contract to ensure global sustainability (Contraction). Hence the phrase 'Contraction and Convergence'.

In the form in which this system was initially presented by Aubrey Meyer, emissions of the past were to be disregarded, with emphasis instead on a fair and sustainable distribution in the present and future. Granted the more-or-less equal needs of present and future people, this disregard for history seemed fair. There were dangers, such as poor countries selling entitlements needed by themselves; but such problems could be overcome by placing a ceiling on emissions-trading. There was also the danger that populations would be boosted to achieve a larger national entitlement or quota; but this too could be remedied by agreeing an early rather than a later cut-off date for population numbers.

However, a scientific discovery has raised new perplexities about the fairness of this system. It turns out that if humanity is to have a 50 per cent chance of avoiding an average temperature rise of more than 2 degrees, its total of carbon emissions has to be limited to an all-time total of one trillion tonnes of carbon. For either a 75 per cent chance of 2 degrees or for a 50 per cent chance of achieving a 1.5 degree ceiling, the limit is some three-quarters of this figure, or 750 billion tonnes. These figures have become known as 'humanity's carbon budget'.

But unfortunately, 55 per cent of the budget of one trillion tonnes had already been emitted by 2009, when these figures were published by Meinhausen and others, and the rest of the budget appeared likely to be emitted by a date early in 2044. An implication was that countries most responsible for the 55 per cent already used up could not fairly claim that their inhabitants should be treated equally with residents of countries with low or negligible historical emissions.

This problem does not of itself undermine Contraction and Convergence, because that system could be modified to involve equal per capita emissions since either 1990 (when it became clear that human emissions were changing the climate) or even since 1750, the approximate date of the onset of the industrial revolution. Aubrey Meyer has adopted an intermediate position.

Yet short of some such modification, some other regimen needs to be found to reduce the entitlements of countries whose wealth derives in large part from historical emissions, so as to permit the development of countries whose comparative poverty corresponds to low emissions in the past, and, significantly, whose residents' basic needs widely remain unsatisfied. These are, after all, largely the countries with the strongest case for compensation for the adverse effects of climate change mostly generated by others.

A different system was proposed by Paul Baer, Tom Athanasiou, Sivan Kartha, and Eric Kemp-Benedict in 2008. Since funding is needed to satisfy unsatisfied basic needs in poor countries as well as for climate change mitigation and adaptation, they advocated recognition of Greenhouse Development Rights, which would be funded by an international tax on everyone with an income greater than the average for Spain. An international authority would deploy the proceeds, financing from this fund mitigation, adaptation, and simultaneously sustainable development for impoverished countries.

Subsequently, following the death of Baer, Athanasiou and Kartha, together with Christian Holz, set about updating their approach, so as to take into account humanity's carbon budget and the 1.5 degree target of the Paris conference. Human rights remained central to their approach; at the same time they have attempted to apportion the sharing of the remaining carbon budget (and the mitigation or carbon and equivalent emissions that it requires) on the bases both of countries' historical responsibility and their economic capacity. The resulting Climate Equity Reference Project seeks to serve the purpose of showing what equity would call for, whether responsibility is confined to the period since 1990 (as seems most reasonable), or is traced back to 1950 or 1850.

Integrated solutions are certainly needed for global problems such as these. But the slender prospect of the world's countries agreeing to any such far-reaching solution, and trusting an international authority with powers of such a scale, make it preferable for these issues not to be addressed in a unified manner, but for climate issues in particular to be tackled at a global level separately from issues of development and related rights. So an adjusted version of Contraction and Convergence, while less radical in its aims and scope, seems preferable.

Yet others have suggested global auctions of entitlements to emit carbon dioxide and other greenhouse gases, trusting market forces to generate the most efficient solution to the problem of distributing such entitlements. But auctions favour countries and companies with the financial muscle to outbid the others, and are unlikely to generate outcomes that could be regarded as just.

In fact, the organizers of the United Nations Paris Climate Conference of December 2015 took the view that no centrally administered system of distribution of entitlements and burdens was likely to prove acceptable, and that it was therefore better for all participant nations to make pre-conference commitments of their own; thereby countries which might not otherwise participate could be induced to do so.

But the aggregated commitments, even if fully implemented, turn out to be insufficient to satisfy the goal (also agreed at that conference) of 2 degrees, let alone that of 1.5 degrees (agreed as preferable), with an expected average temperature rise of towards 3 degrees being more likely. Yet provision was also made for these commitments to be reconsidered in the course of regular reviews, and this provision gives hope that they may be revised enough (and soon enough) for the agreed target limits to temperature increase to be attained.

It is too early to say whether the Paris Agreement can sufficiently alleviate climate change. A key element consists in its provision for regular reviews, in which national commitments can be stepped up. It this takes place, it is still not too late to mitigate average temperature increases and limit them to the agreed level of 1.5 degrees.

Meanwhile, the decision of President Trump (2017) to withdraw the USA from the agreement could prove disastrous, although many American states and corporations are maintaining their own commitment, and it seems possible that the USA might return to participation before its withdrawal takes effect. American readers are encouraged to use their voices (and if necessary their votes) to secure a US return to an agreement that, for all its deficiencies, is an indispensable step towards resolving the greatest global environmental problem of our time. American commitment to yet deeper emissions cuts will be vital across the coming decades.

Collectively the various national commitments mean that the agreement is not too far removed from one that is equitable; while the participation of growing economies such as those of China and India means that it has the potential to change the world definitively. A major criticism is that no mention whatever was made in the agreement of compensation, despite the harms inflicted on less developed countries by the emissions of richer countries whose prosperity has been achieved through historical emissions, many of them emitted since it became recognized that greenhouse gases were disrupting the global climate.

But as long as adequate international provision is made for assisting the adaptation of the poorer countries, this provision may play to some degree the role that compensation would have played, albeit unacknowledged. At the same time, more vigorous forms of mitigation could still prevent the submersion beneath the ocean of small islands and their settlements, for which no envisageable form of compensation could provide redress.

Reducing greenhouse gas (and other) emissions

To prevent climate catastrophe, greenhouse gas emissions need to be eliminated. This is not a hopeless project. For example, emissions of carbon dioxide (as opposed to other greenhouse gases) appear to have stabilized in 2016, though this claim is contested. Meanwhile the Kigali conference of the same year agreed to curtail emissions of HFCs, a greenhouse gas so potent that its emissions have vastly more impact per unit than carbon dioxide. This agreement is expected to arrest the rise of average temperatures by as much as 0.5 degrees in the course of the current century. However, if the permafrost of Siberia continues to melt, there is a danger of gargantuan and potent emissions of methane (currently buried there) raising greenhouse gas levels yet further.

Thus much more needs to be done. In particular, electricity generation based on coal, oil, and gas needs to be replaced with electricity from renewables. In theory, the introduction of Carbon Capture and Storage (CCS) could forestall the need for this replacement. But CCS technology is far from guaranteed to work, and leaks from underground storage of carbon dioxide would completely undermine the expected benefits. Gas-fired generation is less detrimental than generation from coal. But because it too produces carbon emissions, it is no solution.

Many advocate replacing generation from fossil fuels with nuclear energy. But the problems involved in safe storage of spent nuclear fuel and in safely decommissioning nuclear power stations remain unresolved; besides, the risk of accidents such as those at Chernobyl and Fukushima cannot be forgotten. It is vital that carbon emissions not be replaced by radioactive emissions. Fortunately, the costs of electricity generation from renewables (solar, wind, hydro, tidal, and wave power) have now fallen below the cost of nuclear energy, and of course these forms of generation are virtually emission-free. Nuclear fusion, also largely emission-free, may eventually supplement renewables, but the necessary technology does not yet exist.

So it is to renewables that we must turn. Their introduction certainly needs to be linked to provision for energy storage for periods when the sun is not shining and the wind is not blowing. But this technology is well-advanced. Precisely which renewable technology a country adopts depends on its geographical situation and technological resources.

In Britain, wind, hydro-electric, and tidal energy are the most appropriate kinds, but solar energy has the greatest potential in warmer places (while already contributing even in Britain, through solar panels on roofs such as my own). In some places, such as Nepal, local hydro-electric generation (unconnected to the national grid) widely offers the greatest prospect of supplying the needs of rural people. In short, renewable energy generation (of one kind or another) needs to be introduced rapidly worldwide.

Meanwhile households in colder countries need to improve their insulation, so as to reduce their energy use. The installation of solar panels, by generating electricity from the sun, also helps reduce the need for electricity generation plants, and is potentially valuable worldwide.

Attention also needs to be focused on vehicles, and curtailing their emissions. At one stage, diesel-powered vehicles were being encouraged by the British government because of their lower carbon output; but their emissions of nitrogen dioxide, nitrous oxide, and air-borne particulates have proved so detrimental to health that diesel engines need to be phased out without delay (rather than by 2040, the target selected by the British government). Automobiles fuelled by petrol must also be phased out and replaced by electric cars, to which automobile manufacturers are already beginning to turn. As long as the electricity with which they are charged is itself generated from renewable sources, this will reduce urban pollution and significantly reduce the carbon footprint of vehicle users.

At the same time, re-afforestation and the planting of trees has an important part to play. The photosynthesis of trees removes carbon dioxide from the atmosphere. Countries that have not yet cut down their forests should be encouraged to preserve them, not least through the kinds of funding agreed at the Nagoya Conference held in Japan in 2010. Logging practices, prevalent in places as far apart as Brazil and Siberia, need either to be curtailed or to be counter-balanced through replanting. Also the reforestation of deforested areas such as Haiti and much of northern Ethiopia (where forests were destroyed through warfare) should be taken forward with energy and resolve, as is currently being done in Cuba.

Many of these changes can only be effected by governments, and on them the central responsibility falls. But much power is in the hands of corporations (such as energy suppliers, car manufacturers, and airlines), which cannot escape their share of responsibility. Individuals and households, however, also have an indispensable role, in setting an example, signing petitions, speaking out, and putting pressure on corporations and governments to play their full part and commit themselves to policies of radical action. Indeed this is a role open to most readers of this book.

Climate engineering

There is increasing discussion of schemes either to reduce carbon dioxide from the atmosphere (carbon dioxide removal or CDR) or to reflect back incoming solar energy (solar radiation management or SRM). Some forms of CDR are innocuous, such as growing more trees (as commended here and in Chapter 5), and painting roofs white to prevent the absorption of solar radiation. There again, soil could be enhanced and carbon kept out of circulation by burying in it biochar (charcoal made from biomass). Or plants could be genetically engineered to increase their carbon capture. Or carbon could be removed from the air by chemical means (direct air capture). Other forms are more questionable, such as saturating the oceans with iron filings so as to encourage the growth of algae (to absorb oceanic carbon dioxide). But this would threaten ocean ecosystems. It would also risk turning the oceans bright green.

The trouble with the more innocuous forms of CDR is that they have a long lead-time, and are unlikely to be effective enough soon enough. This is what drives some technologists either to more radical forms of CDR, or to placing sunlight-reflecting aerosols in the stratosphere, a form of SRM.

Installing such aerosols was first proposed as a measure to supplement mitigation, but is occasionally suggested as itself a solution to the problem of increasing carbon emissions. It would be comparatively cheap, and could be initiated by a single state without waiting for global agreement. However, sulphur aerosols could acidify the atmosphere, and exacerbate the acidification of the oceans (which is already a problem). Also, once the process of installation had begun, suspending it would lead to rapid increases of carbon concentrations in the atmosphere, and so it might have to be continued indefinitely. There again, the colour of the sky might change for ever.

Another suggested form of SRM is 'marine cloud brightening', in which oceanic clouds would reflect back more sunlight than at present, after treatment with materials such as sea salt. But this process could generate further climate change, not least through increasing rainfall. We should probably hesitate before risking changes to (for example) the cycle of monsoons.

Accordingly, climate engineering, except in the innocuous forms of tree-planting and roof-painting, should be avoided if at all possible. The risk that one or another power will initiate it forms yet another reason why the existing global agreement on mitigation should be made as effective as possible as soon as possible, to forestall the temptation to put into effect any such technological fix.

Grounds for climate action

While action on climate change can readily be grounded in the self-interest of the current generation, daily assaulted by air-borne pollution as it is, there are multiple further grounds. As I write, many people are being driven into exile from sea-level rise, while increasing numbers have become victims of hurricanes, wildfires, and droughts of increasing frequency and intensity. Nor can we expect respite from these alarming tendencies. If too little is done, the next generation will have to undergo yet worse disruption, and their successors worse still. So even readers with an anthropocentric stance (Social Ecologists included) have every reason to participate in campaigns for climate action.

Biocentrists, however, have additional grounds, just as they have grounds for further kinds of species conservation. Climate change is disrupting ecosystems and driving species to extinction at an alarming rate. Some of these are species capable of outliving humanity on our planet, if not driven to extinction in coming decades; so climate change is actually affecting the human legacy to planetary life of the post-human future.

Meanwhile climate change strikes at species useful to humanity and at ones lacking such instrumental value, but whose continued flourishing has a value of its own. Planetary stewards are obliged to care for the continued life of such species, if only through policies designed to allow them to live their own lives unsullied by anthropogenic impacts. It is possible that continuing climate change would allow new species to evolve; but the widespread decimation of existing species, and the risks of this process worsening, strongly adds to the biocentric grounds for climate action, as it also does for habitat preservation.

Yet it might be objected that both the Contraction-and-Convergence approach and its rivals are unduly human-centred. For non-human creatures obviously need to use the atmosphere as well as humans. The emissions of domestic animals (including farm animals) clearly have to be included among the emission entitlements of their human owners, and should not be forgotten. But wild animals must not be disregarded either.

Schemes for mitigation clearly need to allow for the emissions of wild animals when permissible totals for human emissions are being calculated. In this way, such schemes can avoid the charge of anthropocentrism. Fortunately the photosynthesis of green plants and trees outweighs the emissions of their respiration; the more trees are planted, the greater the prospects are for alleviating climate change. While biocentrists can welcome this contribution, everyone needs to take it into account.

Since the Paris agreement relies on national commitments, these will each need to allow for the contributions of the wildlife of the relevant country, and of predictable changes from trends in farming and forestry. The national contributions of richer countries should also include subsidies paid to preserve the forests and wetlands of developing countries. For on such contributions both the future of biodiversity and the role of remaining forests in preventing increases in atmospheric carbon depend.

Climate change as a test-case for environmental ethics

Although the founders of environmental ethics did not envisage the kind of anthropogenic climate change discovered around 1990, climate change epitomizes themes they introduced. Thus

Richard Routley's proposal of a non-anthropocentric ethic and his writings about the dangers to future generations of nuclear energy generation are echoed in the concerns of current environmental ethics to mitigate climate change for the sake of non-human kinds and future generations.

There again, the advocacy of Arne Naess to heed the needs of developing countries, future people, and non-human species has become mainstream within contemporary environmental concern. Similarly, Holmes Rolston's grounding of ecological 'oughts' in the value of healthy ecosystems is ever more widely endorsed wherever concern to preserve forests, wetlands, estuaries, and reefs from pollution and climate change is articulated.

Likewise climate change illustrates the way in which neither nature nor the environment can be considered as given, let alone static. While their changing systems are affected by human action, human life itself remains dependent on their relatively intact functioning. And even if it were ever possible to disregard future generations, climate change makes such disregard not merely imprudent but positively perverse. Action to rescue the environment from degradation and pollution (such as global warming) is required by ethical principles, moral virtues, and the promotion of the best available outcomes alike.

Both sustainable development and ecological preservation depend on strong action, both individual and governmental, local and global, in matters of climate change. Despite their disagreements, Deep Ecology, ecofeminism, Social Ecology, the environmental justice movement, and Green parties can (and must) unite in support of such action. Jewish, Christian, and Islamic supporters of stewardship, whether anthropocentric or not, need to lend such action their support, as do adherents of secular understandings of stewardship, and the adherents of other religions seeking to preserve the Earth and its sacred places. For the future of the planet and all its species is at stake.

References

Chapter 1: Origins

Rachel Carson, *Silent Spring* (London: Hamish Hamilton, 1962).

Clarence J. Glacken, *Traces on the Rhodian Shore* (Berkeley, CA: University of California Press, 1967).

Kenneth E. Goodpaster, 'On Being Morally Considerable', *Journal of Philosophy* 75 (1978): 308–25.

Aldo Leopold, *A Sand County Almanac* (New York: Oxford University Press, 1949).

Arne Naess, 'The Shallow and the Deep, Long-Range Ecology Movement. A Summary', *Inquiry* 16 (1973): 95–100.

Bryan G. Norton, *Sustainability: A Philosophy of Adaptive Ecosystem Management* (Chicago: University of Chicago Press, 2005).

John Passmore, *Man's Responsibility for Nature* (London: Duckworth, 1974 and 1980).

Holmes Rolston III, 'Is There an Ecological Ethic?', *Ethics* 85 (1975): 93–109.

Richard Routley (later Sylvan), 'Is There a Need for a New, an Environmental Ethic?', *Proceedings of the World Congress of Philosophy* (Varna: World Congress of Philosophy, 1973), 205–10.

Chapter 2: Some key concepts

Carmen Velayos Castelo, 'Reflections on Stoic Logocentrism,' *Environmental Ethics* 18 (1996): 291–96.

Nigel Dower, 'The Idea of the Environment', in Robin Attfield and Andrew Belsey (eds), *Philosophy and the Natural Environment* (Cambridge: Cambridge University Press, 1964), 143–56.

Kenneth E. Goodpaster, 'On Being Morally Considerable', *Journal of Philosophy* 75 (1978): 308–25.

Alan Holland, 'Fortitude and Tragedy: the Prospects for a Stoic Environmentalism', in Laura Westra and Thomas M. Robinson (eds), *The Greeks and the Environment* (Lanham, MD: Rowman & Littlefield, 1997), 151–66.

Emma Marris, *Rambunctious Garden: Saving Nature in a Post-wild World* (New York and London: Bloomsbury, 2011), 4–6.

John Stuart Mill, 'Nature', *Three Essays on Religion* (New York: Greenwood Press, 1969), 3–65.

Montreal Protocol on Substances that Deplete the Ozone Layer (1987): http://www.ciesin.org/TG/PI/POLICY/montpro.html (accessed 20 February 2018).

Peter Singer, *Practical Ethics*, 2nd edn (Cambridge: Cambridge University Press, 1993).

James P. Sterba, 'A Biocentrist Strikes Back', *Environmental Ethics* 20 (1998): 361–76.

Paul Taylor, *Respect for Nature: A Theory of Environmental Ethics* (Princeton: Princeton University Press, 1986).

Chapter 3: Future generations

Declaration on the Responsibilities of the Present Generations Towards Future Generations (1997): http://portal.unesco.org/en/ev.php-URL_ID=13178&URL_DO=DO_TOPIC&URL_SECTION=201.html (accessed 20 February 2018).

Kristian Skagan Ekeli, 'Giving a Voice to Posterity—Deliberative Democracy and Representation of Future People', in Robin Attfield (ed.), *The Ethics of the Environment* (Farnham: Ashgate, 2008 [2005]), 499–520.

Hilary Graham, J. Martin Bland, Richard Cookson, Mona Kanaan, and Piran C. L. White, 'Does the Public Favour Policies that Protect Future Generations: Evidence from a British Survey of Adults', *Journal of Social Policy* 46/3 (July 2017): 423–45.

Kigali Agreement: https://www.clearias.com/kigali-agreement/ (accessed 20 February 2018).

Onora O'Neill, *Towards Justice and Virtue* (Cambridge: Cambridge University Press, 1996).

Derek Parfit, *Reasons and Persons* (Oxford: Clarendon Press, 1984).

Christopher Stone, *Should Trees Have Standing?* (Los Altos, CA: William Kaufman, 1974).

Thomas H. Thompson, 'Are We Obligated to Future Others?', in Ernest Partridge (ed.), *Responsibilities to Future Generations: Environmental Ethics* (Buffalo, NY: Prometheus, 1981), 195–202.

Edward O. Wilson, 'Biophilia and the Conservation Ethic', in Stephen R. Kellert and Edward O. Wilson (eds), *The Biophilia Hypothesis* (Washington, DC: Island Press, 1993), 31–41.

World Future Council, *National Policies & International Instruments to Protect the Rights of Future Generations: A Legal Research Paper* (Hamburg: World Future Council, 2009).

Chapter 4: Principles for right action

Aristotle, *Nicomachean Ethics*, trans. and ed. Roger Crisp (Cambridge: Cambridge University Press, 2000).

Robin Attfield, *Ethics: An Overview* (London: Continuum/Bloomsbury, 2012).

Seyla Benhabib, *Situating the Self: Gender, Community and Post-Modernism in Contemporary Ethics* (Routledge: New York, 1992).

Rosalind Hursthouse, *On Virtue Ethics* (Oxford: Oxford University Press, 1999).

Dale Jamieson, 'When Utilitarians Should Be Virtue Theorists', *Utilitas* 19/2 (2007): 160–83.

Immanuel Kant, *The Moral Law: Kant's Groundwork of the Metaphysics of Morals*, trans. H.J. Paton (1948) (London, Routledge, 2005 [1785]).

John Rawls, *A Theory of Justice* (Cambridge, MA: Harvard University Press, 1971).

John Rawls, *Political Liberalism* (New York: Columbia University Press, 1993).

Holmes Rolston III, 'Environmental Virtue Ethics: Half the Truth, but Dangerous as a Whole', in Ronald Sandler and Philip Cafaro (eds.), *Environmental Virtue Ethics* (Lanham, MD: Rowman & Littlefield, 2005), 61–78.

Paul W. Taylor, *Respect for Nature: A Theory of Environmental Ethics* (Princeton: Princeton University Press, 1986).

Chapter 5: Sustainability and preservation

Wilfred Beckerman, 'Sustainable Development: Is It a Useful Concept?', *Environmental Values*, 3/3 (1994): 191–204.

The Convention on Biological Diversity (1992): https://www.cbd.int/doc/legal/cbd-en.pdf (accessed 23 February 2018).

Herman E. Daly (ed.), *Toward a Steady-State Economy* (San Francisco: W.H. Freeman, 1973).

Herman E. Daly, 'On Wilfred Beckerman's Critique of Sustainable Development', *Environmental Values*, 4/1 (1995): 49–55.

Anne E. Ehrlich and Paul Ehrlich, 'Extinction: Life in Peril', in Lori Gruen and Dale Jamieson (eds), *Reflecting on Nature: Readings in Environmental Philosophy* (New York: Oxford University Press, 1994), 335–42.

The Nagoya Protocol on Access and Benefit-Sharing (2010): https://www.cbd.int/abs/ (accessed 23 February 2018).

Rio Declaration on Environment and Development (1992): http://www.un.org/documents/ga/conf151/aconf15126-1annex1.htm (accessed 23 February 2018).

United Nations, 'Millennium Development Goals' (New York: United Nations, 2000): http://www.un.org/millenniumgoals/bkgd.shtml (accessed 7 April 2017).

United Nations, 'Sustainable Development Goals: 17 Goals to Transform Our World' (New York: United Nations, 2015): http://www.un.org/sustainabledevelopment/sustainable-development-goals/ (accessed 11 April 2017).

World Commission on Environment and Development, *Our Common Future* ('The Brundtland Report') (Oxford: Oxford University Press, 1987).

Chapter 6: Social and political movements

Marion Hourdequin, *Environmental Ethics: From Theory to Practice* (London: Bloomsbury, 2015).

Workineh Kelbessa, 'Environmental Injustice in Africa', *Contemporary Pragmatism* 9/1 (2012): 99–132.

Carolyn Merchant, *The Death of Nature, Women, Ecology and the Scientific Revolution* (San Francisco: HarperSanFrancisco, 1980).

Mary Midgley, *Beast and Man: The Roots of Human Nature* (Hassocks: Harvester Press, 1979).

Arne Naess, 'The Shallow and the Deep, Long-Range Ecology Movement. A Summary', *Inquiry* 16 (1973): 95–100.

Konrad Ott, 'Variants of De-growth and Deliberative Democracy: A Habermasian Proposal', *Futures* 44 (2012): 571–81.

Val Plumwood, 'Nature, Self and Gender: Feminism, Environmental Philosophy and the Critique of Rationalism', *Hypatia* 6 (1991): 3–27.

The Principles of Environmental Justice (adopted by First National People of Color Environmental Leadership Summit, Washington, DC, 1991): https://www.nrdc.org/resources/principles-environmental-justice-ej (accessed 26 February 2018).

James Sterba, *Justice for Here and Now* (New York: Cambridge University Press, 1998).

Karen Warren, 'The Power and Promise of Ecological Feminism', *Environmental Ethics* 12 (1990): 121–46.

Chapter 7: Environmental ethics and religion

Susan Power Bratton, 'The Original Desert Solitaire: Early Christian Monasticism and Wilderness', *Environmental Ethics* 10 (1988): 31–53.

Haudenosaunee (Iroquois): http://www.peacecouncil.net/NOON/articles/culture1.html (accessed 19 February 2018).

S. Nomanul Haq, 'Islam', in Dale Jamieson (ed.), *A Companion to Environmental Philosophy* (Malden, MA: Blackwell, 2001), 111–29.

Jonathan Helfand, 'The Earth is the Lord's: Judaism and Environmental Ethics', in Eugene C. Hargrove (ed.), *Religion and Environmental Ethics* (Athens, GA: University of Georgia Press, 1986), 38–52.

Simon P. James, *Environmental Philosophy: An Introduction* (Cambridge: Polity, 2015).

Fazlun Khalid, 'The Disconnected People', in Fazlun Khalid and Joanne O'Brien (eds), *Islam and Ecology* (London: Cassell, 1992), 99–111.

James Lovelock, 'The Fallible Concept of Stewardship of the Earth', in R.J. Berry (ed.), *Environmental Stewardship* (London: T. and T. Clark, 2006), 106–11.

Clare Palmer, 'Stewardship: A Case Study in Environmental Ethics', in R.J. Berry (ed.), *Environmental Stewardship* (London: T. and T. Clark, 2006), 63–75.

Jennifer Welchman, 'A Defence of Environmental Stewardship', *Environmental Values* 21/3 (2012): 297–316.

Lynn White, Jr, 'The Historical Roots of Our Ecologic Crisis', *Science* 155/37 (1967): 1203–7.

Richard Worrell and Michael C. Appleby, 'Stewardship of Natural Resources: Definition, Ethical and Practical Aspects', *Journal of Agricultural and Environmental Ethics* 12 (2000): 263–77.

Chapter 8: The ethics of climate change

Carbon Engineering, 'About Direct Air Capture': http://carbonengineering.com/about-dac/ (accessed 1 March 2018).

Committee on Geoengineering Climate, Board of Atmospheric Sciences and Climate, Ocean Studies Board, Division on Earth and Life Studies, and National Research Council, *Climate Intervention: Reducing Sunlight to Cool Earth* (Washington, DC: National Academies Press, 2015).

Kirstin Dow and Thomas E. Downing, *The Atlas of Climate Change*, 3rd edn (Brighton: Earthscan, 2011).

Christian Holz, Sivan Kartha, and Tom Athanasiou, *Climate Equity Reference Project*: https://climateequityreference.org/ (accessed 27 September 2017).

Intergovernmental Panel on Climate Change, 'Climate Change 2013: The Physical Science Basis', Fifth Assessment Report: http://www.climatechange2013.org/images/report/WG1AR5_ALL_FINAL.pdf (accessed 1 March 2018).

International Biochar Initiative: https://www.biochar-international.org/ (accessed 4 July 2018).

Malte Meinhausen et al., 'Greenhouse Gas Emission Targets for Limiting Global Warming to 2° C', *Nature* 458 (30 April 2009): 1158–63.

Aubrey Meyer, *Contraction & Convergence: The Global Solution to Climate Change: Schumacher Briefing No. 5* (Totnes: Green Books, 2005).

Peter Singer, *One World: The Ethics of Globalization*, 2nd edn (New Haven, CT: Yale University Press, 2002).

United Nations Paris Agreement: http://unfccc.int/paris_agreement/items/9485.php (accessed 1 March 2018).

Further reading

Chapter 1: Origins

Robin Attfield, *The Ethics of Environmental Concern*, 2nd edn (Athens, GA: University of Georgia Press, 1991 [1983]).

Hans Jonas, *The Imperative of Responsibility*, trans. Hans Jonas and David Herr (Chicago: University of Chicago Press, 1984).

George Perkins Marsh, *Man and Nature*, ed. David Lowenthal (Seattle: University of Washington Press, 2003 [1864]).

Chapter 2: Some key concepts

Robin Attfield, 'Climate Change, Environmental Ethics, and Biocentrism', in Ved Nanda (ed.), *Climate Change and Environmental Ethics* (New Brunswick, NJ: Transaction, 2011), 31–41.

Robin Attfield, *Environmental Ethics: An Overview for the Twenty-First Century*, 2nd edn (Cambridge: Polity, 2014).

John M. Rist, *Stoic Philosophy* (London: Cambridge University Press, 2010 [1969]).

Holmes Rolston III, 'Can and Ought We to Follow Nature?', *Environmental Ethics* 1 (1979): 7–30.

Holmes Rolston, III, *Genes Genesis and God: Values and Their Origins in Natural and Human History* (Cambridge: Cambridge University Press, 1999).

Lars Samuelsson, 'Reasons and Values in Environmental Ethics', *Environmental Values* 19 (2010) 517–35.

G.J. Warnock, *The Object of Morality* (New York: Methuen, 1971).

Chapter 3: Future generations

Joseph Addison, *The Spectator* 20 August 1714: 583.

Robin Attfield, 'Future Generations', in Hen ten Have (ed.), *Encyclopedia of Global Bioethics*, 3 vols (Cham, Switzerland: Springer, 2016).

Martin Hughes-Games, 'Why Planet Earth II Should Have Been Taxed', *The Guardian* 2 January 2017: 27.

Workineh Kelbessa, 'Can African Environmental Ethics Contribute to Environmental Policy in Africa?', *Environmental Ethics* 36 (2014): 31–61.

Hans Rosling, Don't Panic—The Truth About Population, https://www. ted.com/playlists/38/hans_rosling_5_talks_on_global_issues (accessed 11 January 2017).

Hans Rosling and Ola Rosling, *Factfulness: Ten Reasons We're Wrong About the World—and Why Things are Better Than You Think* (London: Hodder & Stoughton, 2018).

Chapter 4: Principles for right action

J. Baird Callicott, 'Animal Liberation: A Triangular Affair', *In Defense of the Land Ethic: Essays in Environmental Philosophy* (Albany, NY: State University of New York, 1989), 15–38.

Brad Hooker, 'The Collapse of Virtue Ethics', *Utilitas* 13.1 (2002): 22–40.

Rosalind Hursthouse, 'Virtue Ethics vs. Rule-Consequentialism: A Reply to Brad Hooker', *Utilitas* 13.1 (2002): 41–53.

W.D. Ross, *The Right and the Good* (Oxford: Clarendon Press, 1930).

Peter Singer, *Animal Liberation: A New Ethic for Our Treatment of Animals* (London: Jonathan Cape, 1976).

World Commission on Environment and Development, *Our Common Future* ('The Brundtland Report') (Oxford: Oxford University Press, 1987).

Chapter 5: Sustainability and preservation

Robin Attfield, 'Sustainability', in Hugh LaFollette (ed.), *International Encyclopedia of Ethics* (Malden, MA: Wiley-Blackwell, 2012). http://www.hughlafollette.com/IEE.htm (accessed 5 July 2018).

Robin Attfield, *The Ethics of the Global Environment*, 2nd edn (Edinburgh: Edinburgh University Press, 2015).

Susan Baker, *Sustainable Development*, 2nd edn (London: Routledge, 2016).

Kenneth E. Boulding, 'The Economics of the Coming Spaceship Earth', in Herman Daly (ed.), *Toward a Steady-State Economy* (San Francisco, W.H. Freeman, 1973), 121–32.

Nicholas Georgescu-Roegen, 'The Entropy Law and the Economic Problem', in Herman Daly (ed.), *Toward a Steady-State Economy* (San Francisco, W.H. Freeman, 1973), 37–49.

Global Goals Campaign: http://www.globalgoals.org (accessed 11 April 2017).

James E. Lovelock, *The Revenge of Gaia: Why the Earth is Fighting Back—and How We Can Still Save Humanity* (London: Penguin, 2006).

Jonathan Watts, 'Destruction of Nature as Dangerous as Climate Change, Scientists Warn', *The Guardian* 23 March 2018. https://www.theguardian.com/environment/2018/mar/23/destruction-of-nature-as-dangerous-as-climate-change-scientists-warn (accessed 24 March 2018).

Chapter 6: Social and political movements

Simone de Beauvoir, *Le Deuxième Sexe* (Paris: Gallimard, 1949; trans. C. Borde and S. Malovany-Chevallier as *The Second Sex* (New York: Alfred A. Knopf, 2010)).

Murray Bookchin, *The Ecology of Freedom* (Montreal: Black Rose Books, 1991).

Warwick Fox, *Towards a Transpersonal Ecology: Developing New Foundations for Environmentalism* (Albany, NY: SUNY Press, 1995).

James E. Lovelock, *Gaia: A New Look at Life on Earth* (Oxford: Oxford University Press, 1979).

Mary Midgley, *Animals and Why They Matter* (Athens, GA: University of Georgia Press, 1983).

Jonathan Porritt, *Seeing Green: The Politics of Ecology Explained* (New York: Basil Blackwell, 1985).

Tom Regan, *The Case for Animal Rights* (London: Routledge & Kegan Paul, 1984).

Kristin Shrader-Frechette, *Environmental Justice: Creating Equality, Reclaiming Democracy* (New York: Oxford University Press, 2002).

Chapter 7: Environmental ethics and religion

R.J. Berry (ed.), *Environmental Stewardship: Critical Perspectives—Past and Present* (London: T & T Clark, 2006).

Black Elk, *The Sacred Pipe*, ed. Joseph Epes Brown (New York: Penguin Books, 1973).

Pope Francis, *Laudato Si'*, 2015. https://laudatosi.com/ (accessed 4 August 2017).

Mogobe B. Ramose, *African Philosophy through Ubuntu*, revised edn (Harare: Mond Books, 2002).

Clarence J. Glacken, *Traces on the Rhodian Shore: Nature and Culture in Western Thought From Ancient Times to the End of the Eighteenth Century* (Berkeley, CA: University of California Press, 1967).

Lynn White Jr, *Medieval Technology and Social Change* (Oxford: Clarendon Press, 1962).

Chapter 8: The ethics of climate change

Donald Brown et al., *White Paper on the Ethical Dimensions of Climate Change* (Philadelphia: Rock Ethics Institute, 2006).

Stephen M. Gardiner, Simon Caney, Dale Jamieson, and Henry Shue (eds), *Climate Ethics: Essential Readings* (Oxford and New York: Oxford University Press, 2010).

Dale Jamieson, 'Climate Change and Global Environmental Justice', in C. A. Miller and P. N. Edwards (eds), *Changing the Atmosphere: Expert Knowledge and Environmental Governance* (Cambridge, MA: MIT Press, 2001).

Genetically engineered carbon capture: https://www.quora.com/Could-we-genetically-modify-plants-to-absorb-a-lot-more-CO2-from-the-atmosphere (accessed 2 March 2018).

Nagoya Protocol on Access and Benefit-sharing: https://www.cbd.int/abs/ (accessed 2 March 2018).

Gernot Wagner and Martin L. Weitzman, *Climate Shock: The Economic Consequences of a Hotter Planet* (Princeton: Princeton University Press, 2015).

Index

CHRISTIAN ETHICS
A Very Short Introduction
D. Stephen Long

This *Very Short Introduction* to Christian ethics introduces the topic by examining its sources and historical basis. D. Stephen Long presents a discussion of the relationship between Christian ethics, modern, and postmodern ethics, and explores practical issues including sex, money, and power. Long recognises the inherent difficulties in bringing together 'Christian' and 'ethics' but argues that this is an important task for both the Christian faith and for ethics. Arguing that Christian ethics are not a precise science, but the cultivation of practical wisdom from a range of sources, Long also discusses some of the failures of the Christian tradition, including the crusades, the conquest, slavery, inquisitions, and the Galileo affair.

www.oup.com/vsi

GEOGRAPHY
A Very Short Introduction
John A. Matthews & David T. Herbert

Modern Geography has come a long way from its historical
roots in exploring foreign lands, and simply mapping and naming
the regions of the world. Spanning both physical and human
Geography, the discipline today is unique as a subject which
can bridge the divide between the sciences and the
humanities, and between the environment and our society.
Using wide-ranging examples from global warming and oil,
to urbanization and ethnicity, this *Very Short Introduction* paints
a broad picture of the current state of Geography, its subject
matter, concepts and methods, and its strengths and
controversies. The book's conclusion is no less than
a manifesto for Geography' future.

'Matthews and Herbert's book is written- as befits the VSI series- in
an accessible prose style and is peppered with attractive and
understandable images, graphs and tables.'

Geographical.

www.oup.com/vsi

GLOBALIZATION
A Very Short Introduction
Manfred Steger

'Globalization' has become one of the defining buzzwords of our time - a term that describes a variety of accelerating economic, political, cultural, ideological, and environmental processes that are rapidly altering our experience of the world. It is by its nature a dynamic topic - and this *Very Short Introduction* has been fully updated for 2009, to include developments in global politics, the impact of terrorism, and environmental issues. Presenting globalization in accessible language as a multifaceted process encompassing global, regional, and local aspects of social life, Manfred B. Steger looks at its causes and effects, examines whether it is a new phenomenon, and explores the question of whether, ultimately, globalization is a good or a bad thing.

www.oup.com/vsi

The European Union
Union
A Very Short Introduction
John Pinder & Simon Usherwood

This *Very Short Introduction* explains the European Union in plain
English. Fully updated for 2007 to include controversial and
current topics such as the Euro currency, the EU's enlargement,
and its role in ongoing world affairs, this accessible guide shows
how and why the EU has developed from 1950 to the present.
Covering a range of topics from the Union's early history and
the ongoing interplay between 'eurosceptics' and federalists, to
the single market, agriculture, and the environment, the authors
examine the successes and failures of the EU, and explain the
choices that lie ahead in the 21st century.

www.oup.com/vsi

Human Rights
A Very Short Introduction
Andrew Clapham

An appeal to human rights in the face of injustice can be a heartfelt and morally justified demand for some, while for others it remains merely an empty slogan. Taking an international perspective and focusing on highly topical issues such as torture, arbitrary detention, privacy, health and discrimination, this *Very Short Introduction* will help readers to understand for themselves the controversies and complexities behind this vitally relevant issue. Looking at the philosophical justification for rights, the historical origins of human rights and how they are formed in law, Andrew Clapham explains what our human rights actually are, what they might be, and where the human rights movement is heading.

www.oup.com/vsi

SOCIAL MEDIA
Very Short Introduction

Join our community
www.oup.com/vsi

- Join us online at the official Very Short Introductions
  Facebook page.
- Access the thoughts and musings of our authors with our
  online **blog**.
- Sign up for our monthly **e-newsletter** to receive information
  on all new titles publishing that month.
- Browse the full range of Very Short Introductions online.
- Read **extracts** from the Introductions for free.
- Visit our library of **Reading Guides**. These guides, written by our
  expert authors will help you to question again, why you think
  what you think.
- If you are a teacher or lecturer you can order inspection
  copies quickly and simply via our website.

ONLINE CATALOGUE
A Very Short Introduction

Our online catalogue is designed to make it easy to find your ideal Very Short Introduction. View the entire collection by subject area, watch author videos, read sample chapters, and download reading guides.